ADAM NICOLSON is a prize-winning
ory, nature and the countryside including
— etaries, *The Gentry* and the acclaimed *The Mighty Dead*. He is
ner of the Royal Society of Literature's Ondaatje Prize, the
Somerset Maugham Award, the W. H. Heinemann Award and
the British Topography Prize. He has written and presented
many television series and lives on a farm in Sussex.

KATE BOXER is a painter and printmaker. She lives and works
in Sussex and London.

Praise for *The Seabird's Cry*:

'Powerful … fascinating … his beautiful and often stark
vignettes of love, loss and cruelty in the world of seabirds
make empathy possible by showing us ways in which their lives
are like our own'   *The Times*

'Nicolson is a fine descriptive writer and he is drawn to
the otherworldly beauty of seabirds … this wonderful book
… will, I think, become a classic of bird literature'

Literary Review

'An elegant and impassioned book'   *Sunday Telegraph*

'Full of fascinating and often gruesome details … in this
compelling book, Nicolson succeeds in expressing his sense of
awe at these magnificent creatures'   *Sunday Times*

ALSO BY ADAM NICOLSON

The National Trust Book of Long Walks

Frontiers

Wetland: Life in the Somerset Levels

Restoration: The Rebuilding of Windsor Castle

Regeneration: The Story of the Dome

Sea Room

God's Secretaries: The Making of the King James Bible

Seamanship: A Voyage Along the Wild Coasts of the British Isles

Men of Honour: Trafalgar and the Making of the English Hero

Earls of Paradise

*Smell of Summer Grass: Pursuing Happiness – Perch Hill
1994–2011*

Sissinghurst: An Unfinished History

Gentry: Six Hundred Years of a Peculiarly English Class

The Mighty Dead: Why Homer Matters

THE
SEABIRD'S
CRY

*The Lives and Loves of Puffins, Gannets
and Other Ocean Voyagers*

ADAM NICOLSON

with birds by Kate Boxer

**WILLIAM
COLLINS**

William Collins
An imprint of HarperCollins*Publishers*
1 London Bridge Street
London SE1 9GF

www.WilliamCollinsBooks.com

First published in Great Britain by William Collins in 2017
This William Collins paperback edition published in 2018

3

Copyright © 2017 Adam Nicolson
Seabird illustrations © Kate Boxer
Maps © John Gilkes

Extracts from 'Squarings' from *Seeing Things* by Seamus Heaney, London, Faber and Faber Ltd, 1991. Reprinted with permission of the publishers and the Estate of Seamus Heaney.

Extract from 'At the Fishhouses' by Elizabeth Bishop from *The Complete Poems 1927–1979*, Farrar, Straus & Giroux, 2008.

Extract from 'The Marks T' Gan By' by Katrina Porteous from *The Lost Music* (Bloodaxe Books, 1996) reproduced with permission of Bloodaxe Books. (www.bloodaxebooks.com)

Adam Nicolson asserts the moral right to be identified
as the author of this work in accordance with the
Copyright, Designs and Patents Act 1988

A catalogue record for this book is
available from the British Library

ISBN 978-0-00-816570-3

Printed and bound in Great Britain by
CPI Group (UK) Ltd, Croydon CR0 4YY

MIX
Paper from
responsible sources
FSC™ C007454

This book is produced from independently certified FSC paper
to ensure responsible forest management.

For more information visit: www.harpercollins.co.uk/green

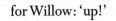

for Willow: 'up!'

Contents

The British Isles

SHETLAND

Foula   Noss

N

········· Flight of Fulmar 1568

Fair Isle

to Charlie-Gibbs
Fracture Zone

Eynhallow
North Rona   *Marwick Head*   ORKNEY
Handa   *Cape Wrath*   Isbister
Butt of Lewis
Lewis
Harris   Shiant
St Kilda   Is.

The Minch

ASSYNT

Hebrides

South Uist
Barra

Aberdeen

*North
Sea*

*North
Atlantic
Ocean*

Tiree
Skerryvore

SCOTLAND

Isle of May
Bass Rock
St Abbs Head

Edinburgh

Farne Islands

from Charlie-Gibbs
Fracture Zone

Tory Island   Rathlin

Ailsa Craig

*NORTHUMBER-
LAND*   Newcastle

Haydon
Bridge

Erris Head

MAYO

Peel

Isle of Man

YORKSHIRE

*Flamborough
Head*

IRELAND

Irish Sea

Aran
Islands   *Galway Bay*

Dublin

LIMERICK

ENGLAND

Blasket
Islands

Swaffham
Elveden

Birmingham

*WORCESTER-
SHIRE*   Cambridge   *SUFFOLK*

Great
Saltee   Skomer
Grassholm   *St George's Channel*
Skokholm

WALES

*PEMBROKE-
SHIRE*   *Gower
Peninsula*

Oxford

London   *R. Thames*

Little
Skellig   Bull
Rock

Cardiff
Bristol

Slimbridge

SURREY   *KENT*

Lundy

Bristol Channel

WESTERN APPROACHES

Clovelly

DEVON

Hastings

Celtic Sea

Penzance   Dartington

Scilly Isles

Land's End   *Lizard*

Start Point

English Channel

FRANCE

| 0 | 50 | 100 | 150 miles |

| 0 | 50 | 100 | 150 | 200 km |

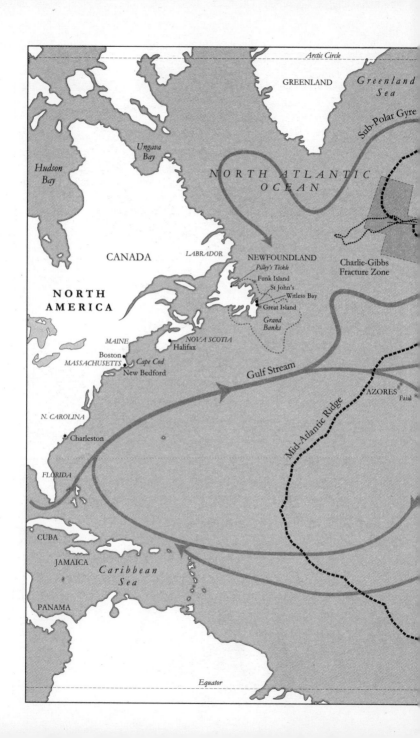

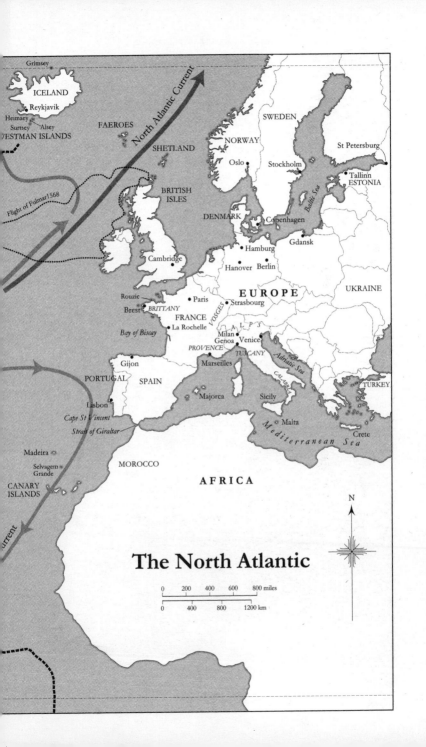

The North Atlantic

Grimsey

ICELAND

Reykjavik

Heimaey
Surtsey · Alsey
WESTMAN ISLANDS

FAEROES

North Atlantic Current

SHETLAND

Flight of Fulmar 1568

BRITISH
ISLES

SWEDEN

NORWAY

Oslo

Stockholm

St Petersburg

Tallinn
ESTONIA

Baltic Sea

DENMARK

Copenhagen

Gdansk

Hamburg

Cambridge

Hanover   Berlin

EUROPE

UKRAINE

Rouzic   · Paris   · Strasbourg

Brest   *BRITTANY*

FRANCE

Bay of Biscay   · La Rochelle

VOSGES   A L P S

Milan
Genoa   Venice

PROVENCE   *TUSCANY*

Adriatic Sea

Gijon

PORTUGAL   SPAIN

Marseilles

CALABRIA

TURKEY

Lisbon

Majorca   Sicily

Cape St Vincent

Strait of Giraltar

Malta

Mediterranean Sea

Crete

Madeira

Selvagem
Grande

CANARY
ISLANDS

MOROCCO

AFRICA

N

Current

0   200   400   600   800 miles

0   400   800   1200 km

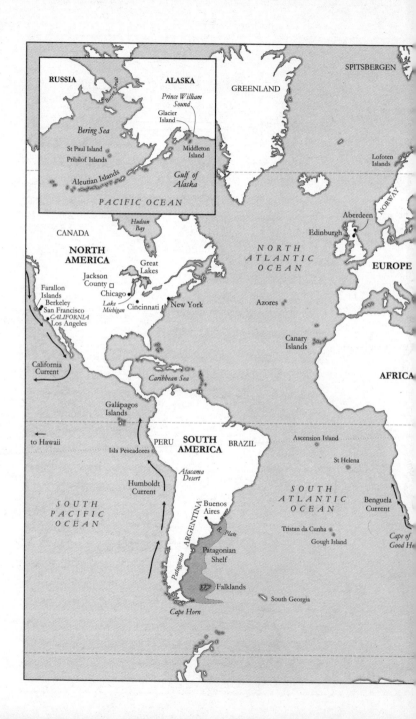

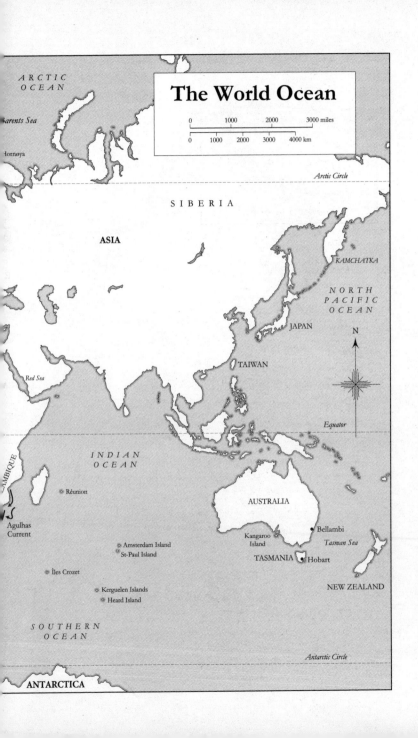

The World Ocean

ARCTIC OCEAN

Barents Sea

Hornøya

Arctic Circle

SIBERIA

ASIA

KAMCHATKA

NORTH
PACIFIC
OCEAN

JAPAN

N

TAIWAN

Red Sea

Equator

INDIAN
OCEAN

MOZAMBIQUE

Réunion

AUSTRALIA

Agulhas
Current

Bellambi

Amsterdam Island
St-Paul Island

Kangaroo
Island

Tasman Sea

TASMANIA

Hobart

Îles Crozet

Kerguelen Islands

NEW ZEALAND

Heard Island

SOUTHERN
OCEAN

Antarctic Circle

ANTARCTICA

| 0 | 1000 | 2000 | 3000 miles |

| 0 | 1000 | 2000 | 3000 | 4000 km |

Introduction

One day, on a steeply sloping grass shelf on the Shiant Isles in the Hebrides, two or three hundred feet above the surface of the Minch, the wide sleeve of sea between the mainland of Scotland and the Outer Isles, I was sitting with Emily Scragg, a young ornithologist who was spending her summer tracking guillemots and razorbills on their fishing trips out from the islands. The sea below us was alight with mid-morning, midsummer sunshine as if polished by it, and we were in shirt sleeves, watching the guillemots on the cliff, dark, handsome birds gathered and jostling there in their thousands. Emily had been putting a GPS tag on one of them, meticulously taped to the feathers on its back, hoping to track it as it foraged in the sea lochs of Lewis and Harris. She had done her work and the bird was back on the rock shelf from which she had picked it ten minutes before. She now had to wait twenty-four hours and the guillemot's data would return.

As we sat there, watching the big roughened crozier-arms of the tide swirling a mile or more out from the headlands below us, a black-backed gull arrived, cruising, easy, sliding low and slow over the guillemot colony, looking for what it might find, and as its shadow crossed them the guillemots in a sudden scare-flight broke away from the cliff, hundreds of them in one dropping, momentous movement, shearing away and down towards the sea. From above, it looked like the rippling of a single wing, a feathered eruption, a dark and magnificent beating of life itself.

Why do you love birds? I asked Emily. Because they fly, she said. That act of release is what is marvellous about them, not as a single done thing but as something that happens again and again, every year, every day, every new life.

The Atlantic seabirds come to breed in places of unremitting hardness. Much of the coastline is a sort of quarry, brutal and intractable, but above it the birds float like beings from the otherworld. They are gravity-free creatures in a place where gravity seems to rule. That, essentially, is what this book is about. Its governing thought is a pair of phrases I read years ago, quoted by Seamus Heaney in one of his lectures as professor of poetry at Oxford. They had been written by the French philosopher and mystic Simone Weil in her collection of aphorisms on grace and transcendence, published after she had died. Weil was exploring the idea that possibility and openness were necessary parts of what was good – the generosity of risk – when she wrote these mind-changing words: '*Obéissance à la pesanteur. Le plus grand péché.*' Obedience to gravity. The greatest sin.

Seabirds never commit it and intuitively, pre-scientifically, we see something oceanic in them, the hint and intimation of another scale of existence, not as part of another, spiritual world, but as the most miraculous and in some ways troubling quality of the one we inhabit. The poets have always understood this. 'I'll be the Bonxie,' Hugh MacDiarmid wrote of the great skua, 'that noble scua,/That infects a'other birds wi' its qualms,' as if one only had to look at a skua to feel the subtlety and edginess of the life it leads.

Seabirds somehow cross the boundary between the matter-of-fact and the imagined. Theirs is the realm both of enlargement and of uncertainty, in which the nature of things is unreliable and in doubt. After both his parents had died within two years of each other in the mid-1980s, Seamus Heaney for a while left behind the poetry of rich and tangible substance and from 'the earth earthy', as Helen Vendler the Harvard critic has called it, quoting St Paul's description of the nature of the first Adam, turned towards a poetry of half-presences and near-absences. Right at the centre of his 1991 collection, which he called *Seeing Things*, are some 'set questions for the ghost of W.B.' – his challenge to Yeats's austere presence over his shoulder.

What came first, the seabird's cry or the soul
Imagined in the dawn cold when it cried?

How habitable is perfected form?
And how inhabited the windy light?

There are no answers, only questions and suggestions, but in that Platonic vision Heaney's imagined soul-seabird is not only the great boundary-crosser, but linked to the emergence and genesis of things. The seabird's cry comes from the beginning of the world.

When any kind of seabirds first materialize in written English, in two eighth-century poems called by later scholars 'The Wanderer' and 'The Seafarer', they are not on the shore but out at sea, referred to by the Anglo-Saxons as 'lone-fliers', inhabiting a strange and ambivalent, half-actual world: half material, half ghostly, half part of our life, half from another realm.

When the friendless man wakes again,
he sees before him the fruitless waves,
sea-birds bathing, wings outspread,
frost and snow half as hail.
Then the heart's wounds deepen, thicken,
sore after sweet – sorrow renewed –
memory of love recurs in the mind;
he greets with open heart, longingly looks
at dead companions. Again they swim away!
Spirits of seamen do not bring
Words you know or songs you love.

Where and what are these seabirds? Are they truly seen by the
grieving and lonely sailor? Has he imagined them? Are they
present on the sea around him? Or hallucinated, figments of
the past now drifting into view? They may be the spirits of his
dead friends but they seem to spread their wings on the cold
of the sea.

I, care-wretched, ice-cold sea,
dwelt in winter on the exile-tracks,
bereaved of friend and without kin,
hung with crystal frost. Hail drove onwards.
I heard only the sea's long moaning,
ice-cold wave. Then the swan's cry
served me for happiness, the call of the gannet
and curlew's weeping, none of our laughter,
sea-gull's mewing, no mead-drink.

The birds' gothic beauty is beyond touching distance. Their
status has long troubled the scholars of these fragmentary
poems, but the point is surely their ambiguity. These are creatures of high latitudes and distant oceans. They thrive in the

sub-polar seas. The further from home we might feel ourselves to be, the more at home they are. That is their world and they are part of what we long for: beauty on the margins of understanding. 'The hiding-places of my power', Wordsworth wrote in *The Prelude*, 'Seem open, I approach, and then they close.' But these birds are more alive than ever in the hiding-places of the north. They are as good imagined – or remembered – as seen, souls and yet not souls, otherness as a dimension of the real. Half-presences, rock-ricochets in their calling. Creatures of the spirit, drenched in ambiguity, half us, half not us, bodies crying in the world.

That, instinctively and subliminally, is what these birds mean to us, voices from the interior of self and ocean, bringing to consciousness those unseen worlds, making apparent what would otherwise be hidden. They are not about transcendence – although they have often been seen as the souls of the dead – but they are, if one can use this expression, an invitation to *inscendence*, a word coined by Thomas Berry (1914–2009), the great modern American philosopher of our relationship to nature. Inscendence does not involve moving beyond the life we know but climbing into it, looking for its kernel, just as MacDiarmid adopts the mad ferocity of the skua. Science, for all that non-scientists disparage it, is dedicated to that urge towards the inward, and the astonishing findings of modern seabird scientists mean that a sense of wonder now emerges not from ignorance of the birds but from understanding them.

In the last couple of decades I have pursued the seabirds across the Atlantic. I have sailed up the west coast of Ireland, to St Kilda, Orkney, Shetland, the Faeroes, Iceland and Norway. I have been to the eastern seaboard of Maine and to Newfoundland, to Ascension, the Falklands, South Georgia, the Canaries and the Azores. The source of this sometimes obsessive fascination with the ways and lives of the birds comes from an experience my father gave me when I was a boy: he first took me when I was eight to the big seabird colony on the Shiant Isles. They are a little cluster of Hebridean islands to which he had been going since he was a student in the 1930s. When he was twenty in 1937, his grandmother had left him some money and, entranced by the idea of remoteness and wildness, he had bought the islands, three small specks of grass and rock, each about a mile long, a total of 500 uninhabited acres, with one rat-ridden bothy, for £1,300.

Adam and Nigel Nicolson on the Shiants, 1971

He loved it there more than anywhere on earth. That was where he went, repeatedly and alone, when on leave from the war in North Africa and Italy. For years he had promised to take me. When at last the day came, after a long journey by train and bus through England and Scotland and then with Gaelic-speaking fishermen and shepherds on a rolling herring-boat across the Minch, I watched carefully as the islands slowly acquired form and contour in front of me. Their grey whale-backed outlines grew and ballooned into something substantial. Dark rocks, grassy slopes, a sheltered bay, the little white house, stony beaches. I had never seen this scale of things before: tall, cliffed, remote, fierce, beautiful, harsh and difficult but, for all that, dazzlingly and almost overwhelmingly thick with the swirl of existence, lichened, the rocks glowing saffron orange on that summer morning, the air and the sea around us filled with 300,000 birds, a pumping, raucous poly-morphous multiversity in which everything was alive and nothing refined.

It was a vision of another world. We landed and picked our way among the colonies. Birds swept over us. We could sit by them and look them in the eye a yard away. Chicks peeped from among the boulders. Puffins growled deep in their burrows. As giant wheels of them turned in the air, their flight feathers rustled and hushed above us. A great black-backed gull swept down and grabbed one in mid-flight. Older victims, stripped of their meat, washed to and fro in the edges of the sea. Beauty and perfection, death, dissolution and life, suffer-ing and triumph: it was all here.

Some people, confronted with a seabird colony like this, turn away from its irreducible presence with a kind of distaste. The multi-layered grab is too much, nightmarish in its half-hidden crevices and suddenness, the shrieking and hawking, the reek of existence. But that inelegance, that dazzling, green-eyed crudity, was the point for me. This was unlike the quiet and careful places I knew at home. It was a section through creation, the column of life itself, drawn down from the eagles a thousand feet above, distant and mesmeric, on through the life and struggle on the cliffs themselves, the stink of ammonia, the chaos of broken shell and kelp stalk, to the lobster rocks and the seal colony below them. Beyond that stretched the open sea from which this life was drawing its sustenance, covered in birds as if paved in them. I saw it as a kind of reality, full-depth, full-intensity, no compromise, the world as it was usually hidden from us. It became a baseline and touchstone for me of what the world might be.

This book, which is an exploration of the ways in which seabirds exert their hold on the human imagination, travels far

Puffins on the boulder screes, the Shiants

beyond the Shiants but has its origins there. I have chosen to describe the lives, habits and destinies of ten birds, or groups of birds. In part they are the ones I know best from the Hebrides, in part the ones I wish were there. These ten divide the sea between them: in their life-habits and body-shapes, their various forms of adaptation, their ways of conquest and triumph. They are the birds which have magnetized my mind, drawn me to them year after year, partly in amazement at the nakedness of their lives, its cruelties and beauties, the undressed nature of their existence, partly in envy, in longing to be what they are.

Each displays a different facet of the central question: how to exist in all three elements. They are the rarest form of creation, the only animals at home on the sea, in the sea, in the air and on land. There are no flying sea mammals, no sea bats, no sea insects, no flying crabs or aerial lobsters. But these birds somehow accommodate – and in fact make glorious – radically differing demands. As flying animals that must breed on land because eggs need to be laid in air if the chick is not to suffocate, seabirds somehow both find the riches and escape the dangers of the sea. How do they do that? And how is it that such grace and power, ingenuity and cleverness, emerge from meeting those demands?

Each has a different answer. The shags and cormorants are, largely, creatures of the coast, rarely adventuring into the depths or the ocean wilds but beautiful, slightly alien, dark-souled and supremely efficient scavengers and divers. They are mostly to be found down near the shore where the gulls, with another set of intriguingly various and often canny solutions to the problem, sit alongside them.

The puffins, guillemots, razorbills and the now extinct great auk – all cousins – are (or were) deep pursuit-divers, plunging in long, hard, wing-driven dives after fast and nutritious prey. These auks mostly occupy the mid-range of the colonies, up from the sea but not in the highest corners, and they fill the ecological niches occupied by penguins in the southern hemisphere. Almost no penguin penetrates north of the Equator and no auk has ever lived south of it, perhaps, as the Canadian bird scientist Anthony Gaston has suggested, because the tropical waters are patrolled by sharks and these birds have never been able to get past that armed and hungry line. Instead, both auks and penguins are confined to the rich cold waters of high latitudes where the sharks could never swim fast enough to catch them.

If the heart of the short-winged auks' lives is under water, the other birds here are the great fliers. The kittiwake is a gull that has abandoned the life of the coast – except when forced to lay its egg there – and taken to the ocean, from whose surface it plucks its food. The gannet is the ferocious dominator of the North Atlantic, an astonishingly powerful plunge-diver and the only bird in that ocean whose numbers and range are expanding under the pressures of modernity. Their nest places also reflect their lives: the kittiwakes' protected on tiny ledges away from any predator, the gannets' in giant and terrifying gatherings of unrivalled ferocity which all other birds and creatures can enter only at the risk of destruction.

The remaining three, the fulmar, the shearwater and the albatross – all related as members of the Procellariiformes, or the Order of the Storm Birds – are the heroes of the story, superbly adapted to life on the ocean, capable of astonishing voyages, long lived, magnificent, thriving on the wind, at home in the turmoil of storm and wave, equipped with all the ease that evolution and their own ability to learn and adapt has given them.

Only 350 out of 11,000-odd species of bird have taken to the sea. For all their differences, a certain way of life unites them, different from most birds: not living a year or two but, in the very oldest albatrosses, up to eighty or ninety years; not raising chicks the season after they are born, but slow to mature, waiting many years before laying an egg; not hoping against hope with eight or nine eggs in each clutch, but often raising a single chick, long incubated in the egg, long fed in the nest; rarely moving on from one partner to the next but often faithful for many years, each parent relying on the other to raise the next generation. These life-histories are shared, significantly, only by the vultures, which must also look for rare concentrations of prey in the wide and hostile sterilities of the world, not at sea but in the desert. These are the edge-choosers, creatures whose lives have stepped beyond the ordinary into environments of such difficulty that they can respond only with a slow, cumulative mastery which amounts in the end to genius.

The writer Arthur Koestler thought our relationship to reality was conducted on three equivalent levels: one informed by the senses; one by the thinking mind; and one by the spirit, the faculty of grasping 'the oceanic', a realm which the senses or the thinking mind couldn't understand, 'just as one could not feel the pull of a magnet with one's skin'. This book is about seabirds, as people have known and are coming to know them, in all three of those dimensions, none given primacy. Each feeds the others and each illuminates the great central fact about them.

> No creature's made so mean,
> But that, some way, it boasts, could we investigate,
> Its supreme worth: fulfils, by ordinance of fate,
> Its momentary task, gets glory all its own,
> Tastes triumph in the world, pre-eminent, alone.

I think of them strung and beaded around the cliffs and crevices of the Atlantic coastlines, from the Skelligs in south-west Ireland to the out-islands in Shetland, the great bird cliffs of the Faeroes, Iceland and Norway, the uncountable thousands on the edge of the Arctic, as the living skin of our ocean shores. They are the florid, rowdy summer clothing of what would otherwise be barren rock. Nearly all of them are in remote places, chosen by the birds for their own protection – or perhaps driven there by man's predatory damage over thousands of years – but they are not a minor presence. They are one of the enormous facts of life from which most of us, most of the time, are kept away. Around the coast of Newfoundland about 35 million seabirds arrive each summer to nest and breed in more than 700 colonies. On tiny Funk Island out in the Atlantic, just over a third of a square mile in extent and one of the last refuges of the great auk, there are, according to the Canadian archaeologist Todd Kristensen, 'more kilograms of edible egg than edible meat in a pod of 100 beluga whales, in a

herd of 800 caribou or in 400 harp seals'. It has been calculated
that each summer, around the shores of the British Isles, 70,000
tons of seabird are on the wing. It is a figure that makes me
laugh, as if they were all one giant bird, weighing half as much
again as Salisbury Cathedral, its feathered wings stretched
across those Atlantic shores like the great bird of dreams.

But they are fragile too, spread thinly across the ocean
world. Seabirds can seem, when you look at them close to and
individually, like victims in the world, almost like refugees,
hopelessly dependent on what life can offer them, subject to
weather and dearth, with failure stalking them at every turn.
But they also have an intriguing doubleness in them: individu-
ated but profoundly collective, individually weak but in their
giant colonies and networks of colonies, a trans-ocean system
of existence, a cumulative assertion of life. That is why they are
one of our imaginative reservoirs, summer ambassadors from
the winter ocean, come to visit us in our mundane existence,
creatures from the otherworld temporarily and for a moment
afloat in ours, and for all their vulnerability a reminder of the
beauty and mystery of existence.

'It is like what we imagine knowledge to be,' Elizabeth
Bishop wrote of the seawater at Lockeport Beach south of
Halifax in Nova Scotia,

> dark, salt, clear, moving, utterly free,
> drawn from the cold hard mouth
> of the world, derived from the rocky breasts
> forever, flowing and drawn, and since
> our knowledge is historical, flowing, and flown.

That fluency and hardness, the cauterizing cold, the oceanic
extent, the taunting inaccessibility, the freedom, the evasive-
ness, the otherness: these are the ingredients of the seabird's

world. And as Bishop wrote in a letter to Robert Lowell, 'Since we do float on an unknown sea I think we should examine the other floating things that come our way carefully; who knows what might depend on it?'

I love the seabirds, partly because I have always met them in the places I love most, the hard and wind-stripped islands and headlands of high latitudes, both north and south, and out at sea in storm and in fair weather. And I have loved them because in those places they have always seemed at home, indifferent to harshness, relishing the conditions, happy to display their beauty in the most demanding moments life can offer. Once in the Faeroes, when I was out with Bjørn Patursson collecting his sheep from the fulmar cliffs on the northern end of Koltur, he asked me what my favourite seabird was. A Manx shear-water, I said, thinking of their effortless flick and cruise, the wafer-thin mastery of air and ocean, the one wingtip feather of the turning bird cutting the sea surface like a knife in the skin.

Ah, yes, Bjørn agreed, savouring the moment, they're delicious roast aren't they? But this is not a book for gourmets, at least of the flesh. It is more about their astonishing lives than about their dead bodies, and draws more from being with the birds and the dogged and persistent observations of the scientists out for year after year in the yowling, stinking colonies than from the kitchen. Carl Safina, the American ecologist, once wrote that the albatross was its own longbow, held taut, and the breeze was its bowstring. That comes near the heart of their beauty: the mutual enveloping of what they are and where they are, with no boundary between organism and environment. Each enfolds the other so that they become the acrobats of ocean and wind, their liquid, floating, commanding presence one aspect of the natural world which requires nothing but a pair of eyes and a readiness to look.

Until now, they have always disappeared over the horizon. People have watched them on their breeding cliffs, glimpsed them from headlands and the decks of ships; but no one knew what they did when they were not there, out at sea. Recoveries of rings from dead birds – and some impressive detective work – began through the twentieth century to establish patterns of migration and dispersal. The information age, largely fuelled by the mobile phone and our hunger for electronics we can carry in our pockets, has changed all that and created a new way of seeing. The satellite loggers, miniaturized heart-monitors, depth-gauges, wetness-detectors and accelerometers, which are all now small enough to attach to living birds, are providing a kind of access to seabird life that we only ever dreamed of before.

Individual life-stories have been pursued for season after season. Journeys are tracked from one pole to the other, from one side of the ocean to the other. Entire seabird lives can now be imagined. The new technology has also begun to reveal idiosyncrasies in seabird behaviour, the differences between individual birds, the choices and inheritances of each of them, the reality of individual family and colony cultures. There is now clear evidence in seabirds of unguessed-at sensitivities to the opportunities and dangers of sea and atmosphere. The seabird body has become the barometer of whole oceans. The levels of stress hormone in individual kittiwakes has been used to measure the abundance of fish in the sea: fewer fish, more stress. Gannets in New Zealand, out foraging for their chicks, fish not only for calories but for specific nutrients to give those chicks a balanced diet. Guillemots can dive to over 600 feet. Shearwaters can smell their way home and fly through a richly scented seascape which guides them to their fishing grounds and back again.

Perhaps only the poets in the past would have thought of seabirds riding the ripples and currents of the world, attuned to how the ocean is, a place of interfolded gifts and threats, but that is what the scientists are seeing now too. Only now have we learned that a wandering albatross flies 5 million miles in its lifetime, only now understood that a kittiwake will leave to breed elsewhere if its close relatives are having difficulty in raising a chick, only now recognize that each puffin holds within its mind a conceptual map of the North Atlantic. The aim of this book, using tradition and science as a kind of twin-pronged tuning fork, is to bring together some of those modern revelations with the older understanding that seabirds are somehow symbolic of the state of ocean and world.

But the owl of Minerva flies at dusk: science is coming to understand the seabirds just as they are dying. By one measure, in the last sixty years they have declined across the world ocean by about two-thirds. The damage threatened by climate change, warmer seas, more acid seas, changing oceanographic patterns, pollution, the effects of industrialized fishing, the loss of habitat and the appetites of the rats and cats we have distributed around the world all ripple through the seabird community like songs to be sung at the apocalypse. The seabirds may have been spirits in the house of myth and legend for as long as human beings have been conscious; they may have lived for 100 million years in their palace of ocean and air, but it looks as though we are now destroying them.

There are counter-trends. Some seabird families are actually expanding in number and in range. There is plenty of evidence of adaptability and resilience in the birds and it would be a mistake to think that, in the light of this grievous damage, we must simply hold our heads in our hands. 'The shape of the creature is the pressure of life against the limit of death,' the art critic Tom Lubbock wrote in his journal in October 2008 as he learned he was dying of a brain tumour. That might be the motto to set over this story. The seabirds push out against

the negative. They are pregnant with meaning and are asser-
tions in a world of denials. They concentrate beauty and coher-
ence. They embody genius. They are the opposite of entropy
and emblems of hope.

At the opening of the Enlightenment, in the mid-seventeenth
century, and in the wake of everything about the separation of
body and soul promulgated by René Descartes, the idea gained
currency that the cries of an animal were no more than the
noises of a machine that was not properly oiled. When in the
1650s a visitor came to Port-Royal, the Jansenist school
outside Paris, at which Pascal was a master and Racine a pupil,
he found the teacher-intellectuals, all of whom were entranced
by Descartes, reflecting on nature.

> There was hardly one of them who wasn't discussing
> *automata* ... Quite unfeelingly they used to hit [their dogs]
> with sticks and ridicule anyone who complained on behalf of
> these animals as if they actually felt any pain. They said they
> were clocks; and the cries they made, when they were hit,
> were only the sound of a little spring which had been
> stretched ... They nailed some poor animals on boards by
> their four feet, and opened them up while they were still
> alive to see the circulation of the blood, an opportunity for
> much conversation.

Descartes himself wanted to know what a beating heart felt
like and cut into a living dog so that he could put his thumb
into the heart muscle as it opened and closed around him. No
one before the seventeenth century thought like that and only
a few think like it now. It is an issue I have found troubling
when researching this book. Several of the most illuminating
experiments and research programmes into the behaviour,
habits and life-structures of the seabirds have involved cruelty.

Birds have had the nerves cut between their brains and their eyes, or between their brains and the organs with which they can detect smells. Others have had different parts of their brains cut away. They have had magnets attached to their heads to see if that confuses their sense of direction. They have been blinded to see if they can still sense the coming of spring and exposed to unnatural regimes of light and dark. Others have had extra eggs put into their nests to see how they cope with the burden. Or extra doses of stress hormone injected into their bloodstream to observe the effects. In laboratory experiments, mercury injected into laying birds has resulted in small and shell-less eggs, malformed embryos, stunted growth in the chicks, shrinking of nerves, strange behaviour and early death. The language used is nearly always euphemistic, never blinded but 'desensitized', never a cut but a 'lesion', never killed but 'sacrificed', as if euphemism could ever make anything better.

These things are rarely mentioned outside the scientific journals in which the results are published and I am aware of my own hypocrisy. The discovery of many of the wonders of seabird behaviour discussed in this book has involved these or equivalent unkindnesses. I am their beneficiary. Nevertheless, large parts of the modern scientific establishment have left behind that Cartesian indifference to the lives and sensibilities of others. For most researchers now, the bird's consciousness and well-being is not an irrelevance, and it is not Descartes but a fascinating and largely neglected figure who now presides over our understandings of nature. Almost no one has heard of him, but Jakob von Uexküll is the Prospero and hidden mage of all modern seabird studies. He was, of all things, a member of the ancient German Baltic nobility, born in his ancestral manor house in Estonia in 1864. His father, a figure out of Tolstoy, was a geologist, deeply loyal to the Tsar, and became in the end the honorary Mayor of Tallin. As a boy, Jakob spent many hours carefully observing beetles, caterpillars and frogs on his father's estate.

When a young man, Uexküll read Kant and became an expert in the body-structures of marine animals, octopuses and sea urchins, making expeditions to the Mediterranean and the Red Sea, investigating the ways in which each organism met, accommodated and understood the world around it. Kant's governing idea, that our minds shape the world we perceive, led Uexküll to his main focus as a

Jakob von Uexküll in 1915 with his son Thure

biologist: understanding the sensory and cognitive structures that shape the perceived environment for each animal species. That idea became central to his world view. He recognized that each organism had what he called its *Umwelt*. The German means 'surrounding world', but more largely, as the primatologist Frans de Waal has described it, Uexküll had in mind a vision of the animal's 'self-centred subjective world, which represents only a small tranche of all available worlds'. Each species lives in its own unique sensory universe, to which we may be partially or wholly blind, and so we must not speak of animal 'cognition' or animal 'intelligence', but in the plural of 'cognitions' and 'intelligences'. Each animal's 'meaning-world' cannot be understood on any terms except its own.

In 1917, the Russian Revolution took Uexküll's estates from him and destroyed any wealth that was tied up in Russian state bonds, which became worthless. The family was now exiled and poor, and Uexküll spent much of his life moving across Europe from one house and one job to another. Adrift in the world, he taught biology to his friend Rainer Maria Rilke and ended up in the 1930s, deeply disapproved of by the Nazis, teaching, among other things, at an institute attached to the aquarium of Hamburg zoo that trained guide-dogs for the blind. Uexküll based his methods on the *Umwelt* concept and devised a way of teaching the dogs that is still in use today.

As the blind man and the dog have two different *Umwelten*, the trainer has to understand the perceptual and cognitive differences between them. A guide-dog, for example, will not mind if a door is only 3 feet high, but the blind man will. The dog's perceptual world has to be extended upwards. They made the trainee dogs walk around a building pulling a cart in which a life-size model of man stood 6 feet tall. The dog soon learned that he had to understand the building in the way a man would understand it. Only by grasping the *Umwelt* of the other species would the dog not be *bedeutungsblind*, 'blind to its significance'.

We are that dog and the rest of creation is the man in the cart. We have to stretch our understanding to accommodate the understanding of others. Konrad Lorenz knew, visited and admired Uexküll. Through Lorenz and later through his follower, admirer and joint recipient of the Nobel prize Niko Tinbergen, Uexküll's legacy has shaped us. The autonomous organism, the creature with an independent and unique meaning-world around it 'like a soap-bubble' as Uexküll said, has become the subject and focus of modern life-sciences. An acknowledgement of the subjectivity of other creatures, until the twentieth century the preserve of poets and dreamers, is now at the heart of any modern scientific understanding of nature. It is a recognition that has yet to spread into other areas of human enterprise. If the claims of nature are considered at all, they are almost always calculated in terms of human benefit. What kind of world, it is often asked, would we be living in if other creatures were absent from it?

The rest of us need to make this switch. Unless we accept the multiplicity of *Umwelten*, with which every creature perceives the world in ways that are unique to it, we will inevitably end up ranking everything on a single scale, and that scale will be measured by our own human standards. If you are stuck in that anthropocentric vision of nature, you will inevitably think that the more like humans any animal seems to be and the more linguistic and even technological skills they seem to have, the cleverer they are. But as Frans de Waal says, 'There are lots of wonderful cognitive adaptations out there that we don't have or need ... Cognitive evolution is marked by many peaks of specialization.' We have no monopoly on intelligence. We would not know how to plunge-dive for herring or locate the Mid-Atlantic Ridge by smell. We could not hang in the updraughts by a cliff or find our way alone all winter across the Atlantic. We would not know how to exist in a form that is not our own. The seabirds are intelligent in ways that are different from ours.

The irony is that *Umwelt* may be integral to all living things. I have watched a golden eagle for an evening displaying to its mate above a kittiwake colony. The kittiwakes were swirling around their cliffs while the eagle was making a series of astonishing folded tuck dives above them, each a plunge through air as driven as a horse displaying to its mares, an arrowhead hunched into the wind, possessing his cube of air, 2 miles wide in each direction, a vast box of wind.

The kittiwakes paid no attention. The eagle was communicating only with another eagle. Its power-ballet, even as it was being hammered again and again by the skuas, the great blackbacks, the ravens and the peregrines, nibbling and nabbing at it as it tried to maintain its dignity, very occasionally rolling over through 180 degrees and showing to its persecutors the talons that can grab a fulmar mid-flight as if it were rubbish or shopping – all of that was quite irrelevant to the kittiwakes, which were untroubled, skirling to and fro above the rocks, shouting at each other, embedded in their world.

It was like two different principles of life in action: the great predator owning the air, the seabirds inhabiting it; one demonstrating its own vastness, the others absorbed in their lives as if nothing existed outside it. I have seen the same thing in among the boulders of an auk colony: thousands of razorbills hawking and juddering around the rocks, big power-birds when you get close to them, equipped with ferocious, striped machete-bills, and between them, in another universe of consciousness, three dark little wrens hopping and peeping in the screes, arguing and shouting over some political or legal question in wren-world, a cockpit 2 feet by 3, indifferent to the armies of black and white giants looming over them.

Why should one be surprised that the eagle and the kitti-wake, the razorbill or the wren, live in different life-spheres? Each bird is wrapped in its unique *Umwelt*, separated from the others by evolution, seeing nothing but the world it sees for itself. If we have through our history been shut into our perceptions, that is probably because enclosure in one's *Umwelt* is a guiding principle of life on earth. And the fact that, led by figures such as Uexküll, Lorenz and Tinbergen, we now seem to be stepping outside that enclosure is a moment of revolu-tionary significance. If *Umwelt* is the product of close and extreme attention to the things that matter in your life, and indifference to those that don't, then we have probably reached the moment when what matters to us has expanded beyond our own narrowly and historically defined interests. For all the vertigo this thought might induce, the human *Umwelt* now is and needs to be global. We have, of necessity, entered the age of empathy.

The airwaves are awash with talk of the Anthropocene, the geological epoch in which the single dominant factor is man. It is a literal truth that every albatross and fulmar has eaten plastic and it is reliably predicted that by 2050, about 99.8 per cent of all seabird species will have plastic in their stomachs. Nevertheless, some foundations have been laid, or at least seedlings planted, for a successor age. The Anthropocene will have brought one geological moment to an end; it could now usher in the Ecozoic, another of Thomas Berry's formulations, a time in which life-forms are understood systematically, when human beings will engage with them not merely as an irrele-vance or ingredients for dinner but as co-actors with us in the *oikos* – the Greek word at the root of ecology and economy – the house of earth. The Ecozoic is life lived in the *oikos*, powered by empathy and enabled by understanding, and this book is, in its way, a manifesto for the Ecozoic, an age which has at its heart the belief that all living beings have a right to life and to the recognition that they have forms of understanding we have never shared and probably never will.

1

Fulmar

I remember as a boy lying down on the very edge of one of the highest cliffs in the Shiants, shuffling up to it on my belly until my nose was out in clean air, the grey cliff, fuzzy with lichen, dropping straight to the surf below, my feet still well back on the grass. There I could watch the fulmars on their own terms, nothing between me and them but the air on which they lived. I followed them for hours with the binoculars as they played in loop after loop in the billowing gusts around us. This was wind-dancing, untroubled and seamless. They slipped in and out of the sunlight, sometimes lit against the dark, then into shadow, a stroboscope of life on the wing. They would come past at head height, a few feet away, the dark observant eye buried in the white of head and neck, a straight and vertiginous look into the consciousness of another animal. Across the northern hemisphere, fulmars can be found in every tone from nearly black to nearly white, but the Shiant birds have always been a wonderful, mottled fusion of white and pale grey, mysteriously dove-like, the first bird that ever made me wonder what its life consisted of.

Do fulmars play? Parts of their brain are full of the neuro-transmitters that are essential for a feeling of reward, and can be flooded with natural opiates, the enkephalins, without which no animal can feel pleasure. Swans have been seen repeatedly surfing on waves off a beach and crows have tobogganed again and again down a snowy Russian rooftop, using a jam-jar lid as their sledge. It is not stupid to think that birds might play, and here from the clifftop it has always looked as if that is what the fulmars were doing: the endless, repeated turns, first on one great circle and then another, skaters outlining discs on the ice, stiff-winged, patient, waiting for the long rotation to take its form, a series of geometries, as if the birds were cutting shapes through the paper of the air.

A fulmar in flight

The air doesn't always comply. Now and then a strange lack of certainty runs through a fulmar, even as it makes these Euclidean diagrams beneath you, a whole-body hesitation, coughing in mid-flight, when it shudders and disassembles, all sleekness gone and all purpose paused, as if waiting for the data stream to resume, which it then does, and the long effortless gestures, milking energy from the wind, continue from one end of the ballroom to the other.

No one knows what that fulmar-fluster is, the sudden gust of stage-fright in the middle of the performance, but it is at least this: a reminder that these creatures are not feathered machines but individuals with idiosyncrasies and incomplete-nesses, immature or senescent, not perfect but animals that are at least perfect enough.

You could spend hours staring at them – I have – and know nothing more at the end than at the beginning, except for an overwhelming sense of the mastery of the birds, the privilege of seeing them like this, not in crisis or ecstasy but in the midst of life, the wonderful ordinariness with which they scarcely distinguish rock from air, life settled on the nest from life on the sustaining wind, as if one were only the other in another form.

James Fisher, the great lover and student of the fulmar, who devoted much of his life to understanding them and their twentieth-century spread across the North Atlantic, consulted Walter Newmark, a glider and sailplane pilot, on the skills of the bird. Newmark was clear: compared with a fulmar, the gulls – the herring gulls, lesser black-backed gulls and kitti-wakes – were frank incompetents. Fulmars, he wrote in *Sailplane and Glider*,

> usually hold their wings stiffer than a gull without the rake forward and sweep back of the wings, relying for fore-and-aft stability on a very well developed tail. The tail muscles are extraordinarily strong, capable of warping the tail up on one side and down on the other, and at the same time twisting the whole assembly and thus putting on bank without using aileron control.

Watch carefully and you will see, as Newmark says, that for all the apparent suavity and continuity of their flight, the bird makes constant tiny adjustments of its wing, raising the inner 'arm' half so that the leading edge is higher than the trailing edge, and lowering the outer 'hand' half the other way, with the leading edge down, or then vice versa, in constant dialogue with the ripples and instability in the wind, soaring on updraughts, diving into the downcurrents, flicking themselves up, splaying that astonishing tail rudder or suddenly on arrival at the nest lowering its legs for full-flap braking. Nothing in three dimensions is beyond this bird.

The crews of old whalers in the Arctic thought fulmars were the spirits of Greenland skippers, somehow released from the burden of their ships and now afloat over the ocean they had previously raided. More than any other northern bird, that sense of boundary-crossing freedom is what they embody on the wing. They come and go without warning, and in the Hebrides, along with other migrants, they were called in Gaelic *eoin shianta*, the eerie or uncanny ones, perhaps because constancy is what one expects of a creature, a material reliability, present if present. But fulmars refuse to belong consistently to the world of which we are a part. The centre of their life is oceanic, beyond any horizon we might know.

Until recently, people have only been able to guess what the fulmars did when they were not to be seen. They were observed at their colonies and by whalers and trawlermen at sea, but the bulk of their ocean life, and more importantly the connection between ocean life and nest life, was effectively invisible. In the last twenty years the modern revolution in remote sensing has changed all that, and no experiment has better demonstrated the miracle and beauty of the fulmar than one conducted on the small uninhabited Orkney island of Eynhallow in 2012.

Eynhallow is the Mecca and Ayers Rock of fulmar studies, as successive generations of the bird have since 1950 been experimented with and prodded there by scientists from the University of Aberdeen, making them one of the most consistently examined populations of wild animals in the world. In May 2012, twenty-two of the Eynhallow fulmars, nesting in the ruins of the abandoned buildings and old field walls, were fitted with GPS loggers, stuck with waterproof tape to the back, on the mantle between the wings. Each bird was also given a geolocator held on to its leg with a cable tie around a ring.

The revelatory fulmar was a big male, number 1568, and was well known to the scientists. He had bred on Eynhallow with the same partner for the previous eleven years and just after midday on 23 May 2012 the Aberdeen scientists grabbed him where he was sitting on his nest at the southern end of Eynhallow, with a view towards Evie on the Orkney mainland. His partner was away fishing. 1568 had already been taking part in the Aberdeen experiments and Ewan Edwards and Paul Thompson removed a geolocator that had been attached to his leg two years previously. With the new GPS and the new geolocator both fitted, 1568 was settled back on his egg, waiting for his partner to return from fishing.

Three days later she came back and at 10.30 that evening 1568 headed out to sea. Until recently, that is where the information would have stopped. Nowadays, with satellites and miniaturization, it is where it begins. Fourteen days later, on their usual morning round of the nests to see which birds had returned, Paul, Ewan and Sian Tarrant, their volunteer field assistant, found 1568 back on his nest. To their excitement, he was still carrying both his tracking devices, which they removed, before putting him back on the nest and then anxiously downloading the data.

Paul Thompson and Ewan Edwards relieve 1568 of his tracking
equipment on Eynhallow, June 2012

That is often the moment when it becomes apparent that the
GPS had stopped working as soon as the bird left the nest. Not
this time: the weather had been calm in late May, with a large
and stable summer high hanging over the British Isles. 'When
the wind blows they go on their travels,' James Fisher had
written. 'Rough weather is as necessary to the fulmar as the
Trade Winds were to the human conquest of the New World.
Rough weather is the fulmar's passport, its transport to
mid-ocean.' 1568 was true to his genes and for two days he
waited for the wind, afloat on the ocean just to the north-west
of Orkney. But then the weather changed. A deep depression
began to build in the central North Atlantic, well to the south
of the waiting bird, and, as it deepened, strong south-easterlies
began to blow. In that wind, he set off to the north-west, a

sustained eleven-hour flight to the channel between Shetland
and the Faeroes, a rich picking ground for the plankton drifting up in the North Atlantic Current. He stayed there almost
a day, hungry from his time on the egg in Eynhallow, now
grazing happily on the meadows of the ocean.

Then the surprise: early in the morning of the fourth day of
his journey, 1568 set off in the wind on a heading of about
250° west-south-west, and flew fast and hard out into the
depths of the North Atlantic for two and half days, a thousand
miles in fifty-five hours. He slowed at night but during the day
sometimes covered more than 40 straight-line miles in an
hour. If you take the zigzag path of his dynamic soaring into
account, he may have been travelling half as fast again. In good
time, with his mate still sitting on their egg in Eynhallow, he
arrived at the destination he had undoubtedly been seeking,
the rich waters around a mountainous and broken section of
the Mid-Atlantic Ridge called the Charlie-Gibbs Fracture
Zone.

There, on a part of the ocean named after the American
Coastguard Weather Station Charlie and the survey ship *Josiah
Willard Gibbs*, two-thirds of the way to Canada, he feasted for
three days, not travelling far, but feeding on the plankton,
squid and fish that gather at that meeting of the warm North
Atlantic Current and the cold fertile waters coming down
from the Arctic. The Charlie-Gibbs Fracture Zone may be
unknown to anyone who is not a marine scientist, but it is one
of the most important places for life in the Atlantic, where
storm petrels and great shearwaters, Cory's shearwaters
coming up from the Azores, even sooty shearwaters from the
Falklands, long-tailed skuas and Arctic terns, along with
turtles, whales, dolphins and vast shoals of migratory fish, all
gather to feed on the density of planktonic life generated in the
nutrient-rich waters churned up from the ocean floor.

The dark eye, squid-catching bill and exterior nostrils of the fulmar

Imagine 1568 out there, surrounded by other birds from Iceland and Canada, from Greenland and far south in the Atlantic, quartering the sea for his prey. He may well be hunting by sight, as that is the role of his large, dark eyes, but like the albatrosses and other tubenosed birds, the fulmar can also certainly smell his prey. This may be a function of distance: the hunting bird may finally see the bait which until then he has only been able to smell. After three days at the Charlie-Gibbs Fracture Zone, during which he had moved gradually westwards so that his furthest point, eight days after he had left the nest, was almost 1,500 miles from Eynhallow, he turned for home but intriguingly did not make a beeline for Orkney, instead flying, now into strong headwinds, to Galway Bay in south-west Ireland. It may be that he was choosing the headwinds that were nearer the centre of the depression and so slightly weaker. If he had gone straight back to Orkney, it would have been a battle against the easterlies on which he had ridden out into the Atlantic.

Arriving in Ireland, he was a long way from home, many hundreds of miles south of Orkney, but the Vikings used to navigate like this: leave the coast of Norway, aim as best you could for the mainland of Britain, hit it somewhere you would recognize and then follow the coast to your original destination. That looks like 1568's method, aiming for the great unmissable wall of Europe to the east and finding it wherever the headwinds had allowed him.

All the same, his geographical understanding was precise. He knew he was to the south of where he needed to be. He could expect that there would be homeward-heading southerlies on the eastern edge of a low, and having fed on the sea for eight hours off the rich sea life in Galway Bay, 1568 turned definitively north along the Atlantic coastline, hugging the shore until he reached the big headland of Erris Head, shoved out into the Atlantic from County Mayo, and from there cut north-east for Tory Island and then the Hebrides. He made his Scottish landfall at the great lighthouse of Skerryvore off the south-west point of Tiree. There again, in the surging tidal overfalls of that difficult stretch of water, he paused and fed for a few hours.

For the one final step on the road home to Eynhallow, 1568 set off from Tiree on the afternoon of his thirteenth day and, clearly reading the geography of north-west Scotland, tracked up the western, Atlantic side of the Outer Hebrides, not the shortest route but the firmest delineation of the meeting of continent and ocean, a sea- and landscape with which he was certainly familiar. From the Butt of Lewis, another of the great corners of Europe's Atlantic façade, he flew to just south of Cape Wrath and from there home to Eynhallow, to the egg and bird that had been his mate for the last ten years. He arrived at nine in the evening on 9 June 2012, having travelled a straight-line distance of nearly 3,900 miles in just over two weeks. After a moment or two together, his mate left for her own (unknown) voyaging and 1568 settled on to the egg, where, as most fulmars do when returning from a foraging trip, he tucked his head under his wing, sitting on the sorrel and the thrift, and slept.

Not every fulmar followed by Paul Thompson and Ewan Edwards lived quite the heroic life of 1568. Some went off to the Norwegian coast, and many fished far more locally, but one thing became clear from a fusion of 1568's journey in May and June, at nesting time, with the other data gathered from the logger he had worn on his leg for the previous two years. He knew what he was doing. 1568 went to the Charlie-Gibbs Fracture Zone because he knew about it. He knew there was a predictable food source there, because he had spent the first few years of his life wandering the oceans exploring and expanding his knowledge, building up a database of where he was likely to find food. So many birds go to the Mid-Atlantic Ridge to fish that he probably followed others there. The Mid-Atlantic Ridge was ingrained in his mind as a source of well-being. His ocean journeys were acts of memory.

Once you hear this story, you cannot watch a fulmar at a summer cliff and see it in the same way. Its aerial ease and swank, its imperial familiarity with the wind, are now revealed only as the outermost tips of an oceanic life. An iceberg of otherness lies hidden beneath that visible surface, and the GPS tracker has seen into those depths in a way no one has ever seen before, just as a creel pulled into a fishing boat brings unsuspected creatures to the light. Here is a bird so attuned to the ways of planet and ocean, not only physically and instinctively but psychologically and even analytically, that it is possible to see in its whole being an intelligence different from but scarcely less than ours. The GPS tracks are a map of that mind, allowing a glimpse into a fulmar's consciousness.

This bird is no aberration: the body of the fulmar has evolved to behave like this. Although they are capable of flapping, they are equipped to soar and glide and even though a puffin is less than half a fulmar's weight, a fulmar's pectoral muscles, which drive flapping flight, are actually smaller than a puffin's. That is the effect of lifestyle on body: a puffin's wings must operate under water, allowing it to propel itself to great depths in pursuit of fish. Long wings wouldn't work under water – the drag would be too much on each stroke – and so the puffin's wings have shrunk to a compromise between wing and fin. The result is that to stay airborne, the puffin can scarcely ever glide, only as it first drops from its cliffside burrow towards the sea. When down there it must beat its wings up to ten times a second, 600 times a minute. To do that, it has developed its large pectoral muscles, that big, puffed-out swelling of the chest which gives the bird its familiar outline and maybe its name.

The fulmar's flying equipment, sustaining its 25-ounce body on 3 feet 6 inches of feather and bone, is lightweight and wonderfully efficient but only when the wind blows. The Scottish ornithologists Bob Furness and David Bryant have investigated the energy budget of fulmars in different wind conditions. If it were calm and the bird had to flap its wings to move all day – they never do, because a calm day is when you will find fulmars sitting on their nests or the sea – they would have to expend 2,000 calories, thirteen times what they would consume sitting or sleeping doing nothing. As the wind increases, life becomes more viable so that when it is blowing at about 20 miles an hour, or Force 4–5, when the sea is just beginning to be covered in breaking crests, they can glide most of the time and the energy they expend on the wing is, astonishingly, as low as what it would be if they were sitting on the nest. At that speed, they still have to beat their wings about once every second, but when the wind notches up a little further, to about 25 miles an hour, or Force 5 to 6, flying costs them nothing. It is the magic horizon. In those winds, fulmars almost never beat their wings at all, but drink in the energy they need from the wind around them. That, and more, was the wind on which 1568 drove more than halfway to Canada.

If you watch a fulmar in a wind that has raised whitecaps all over the surface of the sea, a place that in a dinghy would be challenging, asking you to reef down and reduce the risk of capsize in a gust, that is when the bird appears most miraculously at home. Fulmars in that kind of breeze are settled, enjoying certainty and control in all the fluid elements around them, at home in a place where body and world interlock. Look at them on days like that, when the wind is blowing through the boundaries of fresh and stiff, and you will see them for what they are: wind-runners, wind-dancers, the wind-spirits, alive with an evolved ability to live with the wind, in it and on it, drawing out its energy to make their own feathered, mobile, ocean-ranging magnificence.

For many other seabirds, including the puffin and the other auks, this relationship is reversed, so that the windier it becomes the more difficult they find it and the more energy they must expend on maintaining their equilibrium. Not the fulmar. There are no whole-body shudders when the wind is pushing towards a gale, but instead a kind of release into dexterity and finesse. They become the virtuosos of wind, deft and perfect not in their existence but in this adroitness, this fit of body to world, as if you were witnessing an organism sliding from dullness into all the gifts and opportunities life in the air might offer. It may even be that that the fulmar's strange all-over shudders, which you see now and then in a calm, are expressions of frustration at windlessness, of the world not being and feeling as the fulmar knows it can be. Nothing but the wind governs where fulmars can live; they are addicts of wind, incapable of living without it, needing wind to animate their lives, and, like most of the albatrosses, prevented from living in the fluky Horse latitudes of the Tropics not from a shortage of prey but because the winds there could not sustain them. In the tropics they would have to flap and would die. High-latitude gales are fulmar life-blood.

*   *   *

I once slept for a night or two in a roofless shieling on North Rona, the most distant of all the British isles, rarely visited except by scientists and shepherds, with only the brutal gannet rock of Sula Sgeir for a near neighbour 10 or 12 miles away to the west. Otherwise, the extraordinarily fertile rolling pastures of North Rona, entirely dependent on the nutrients brought up from the sea and spattered all over it by the birds, are separated from Cape Wrath and the Butt of Lewis by at least 40 often wild miles of the North Atlantic. It is the nearest one can come in Europe to sleeping on the ocean, your ears full all night of the sea churning, and it is a place for the fulmars. They nest in the ruined buildings and in the morning as I lay in my bivvy bag inside the broken walls, they were flying over me no more than 2 feet above my head, unaware of any human presence and emitting the most beautiful sound I have ever heard a seabird make: not a cry but the rustling of air in their feathers, a kind of folded aural blanket overhead, just up from silent, moth- and owl-like, breathy, as if steam were venting from a chimney far away down a valley, or air was somehow being blown through the layers of a mattress, every edge of the sound softened and feathered, the fulmars making their own wind-music within touching distance of my hands.

That is not the sound you will usually hear from a fulmar at its nest. After an autumn away, they are back at their cliffs in December, but only stay there during calms. When the wind blows, they are off out in the Atlantic, tacking and sliding along the waves, until April when serious recourting of their partners begins. That is the moment for the fulmar's song, although no word could be less appropriate to a sound that then lasts all summer, and is indistinguishable whether uttered in love or hatred. Over the centuries, everyone has struggled to characterize it. Seventeenth-century Dutch whalers thought fulmars sounded like frogs chirping in the distance (James Fisher's favourite description). Others have thought them shrill, worrying, quarrelling, like 'a hen calling her chickens', chuckling, somewhere 'between the cackle of a hen and the quack of a duck', 'like the voice of a man in the deepest agony', 'a complacent croak', clattering, jabbering, scolding, a staccato quack-cawing, a retching.

James Fisher, who over his years of devotion to the fulmar gathered these versions of their cry, had his own pair of wonderful descriptions. Listening to them whirring and clucking all round him on the giant cliffs of St Kilda, he thought they sounded at a distance 'like the fizzing, whizzing noise, the crescendo of clockwork, that goes on inside a grandfather clock just before it strikes'. But, nearer to, it was more impassioned, 'the fantastic, ecstatic sound that a trio of fulmars, displaying on or "discussing" the ownership of a nest site could produce', which Fisher transcribed as:

aaark-ag-ah-ak-arrr-brrae-buck-caw-cok-coo-ek-ga-gä-gaggeragaggagagga-gagagagaga-gah-gerr-grorr-grrr-ha-haw-hough-i-kaka-karo-kekerek-kertchük-kokokok-kraw-kuh-kurr-kwakwakwa-oh-ok-ork-r-rherrrrr-roo-tchück-uä-ug-uh-uk-urg-urk-wib

Carol Ann Duffy, thinking of a wonderful night of life-expanding passion 'remembered hearing, clearly/but distantly, a siren some streets away – *de/da de da de da* – which mingled with my own/absurd cries, so that I looked up, even then, to see my fingers counting themselves, dancing'. Sometimes looking at the fulmar's gaping, mouth-opening, mutually frenzied, head-bobbing, nibbling, shout-laughing version of 'I love you/I hate you', which can begin calmly enough but then builds to a tumultuous, guffawing, totalizing climax, I have thought that only in the ecstatic moments of life can we come near to knowing the reality of a bird's mind.

They mate in April but then leave. There is no advantage in staying together at the nest, and the Eynhallow fulmars, fitted with their daylight-recorders, fly either out into the Atlantic or so far north that the recorders become useless and during this 'pre-laying exodus' the scientists lose track of them. The birds are ranging somewhere in the ocean north of Norway, where by early spring daylight is already constant and the birds are living in unbroken light. They return and the single egg is laid in May. Even with their giant oceanic absences, the birds share the incubation until the chick is hatched, usually soon after Midsummer's Day. They must then look after it together for nearly two months before it is capable of flight. To begin with, each takes turns to provision it from the ocean. They cannot, as near-shore fishers such as puffins or guillemots do, bring back whole fish from their distant foraging and so the fulmars, like the albatrosses, petrels and other tubenosed birds, have evolved a method of preserving nutrients in their own gut, saving a rich foul-smelling oil in the folds of their stomach. On returning to the nest, the chick puts its bill inside the parent's mouth and greedily absorbs the oil that is regurgitated for it. It is another adaptation to a life that draws its sustenance from the ocean but must return to land to bring up the next generation, the two poles of existence which a fulmar must swing between. This sticky, stinking oil which gave the fulmar its Viking name – it means 'foul gull' in Old Norse – is also used by the bird in defence, spurted out two or three times at any threatening predator, clinging to the feathers of any gull or bird of prey and sometimes making them incapable of flight. Even very young fulmars project their vomit like this – you do not want to be on the end of it – and James Fisher claimed he had heard of one chick vomiting at a man when still inside the egg, through a gap in the shell it had just started to break open.

To begin with, for about a fortnight, at least one parent stays with the fluffy white puffball of the chick, topping it up with little oil meals now and then, pushing it urgently towards the state where it can survive without a daily meal. Its ability to chuck foul and feather-clinging oil over any threat means both parents can leave it as soon as it is big and fat enough to survive a little neglect. After two weeks or so, from early July onwards, when the chicks have reached a certain size, the tufty fulmar nests on the cliffs are often occupied only by one innocent-looking bundle of soft white down. Dark eyes and a sharp black bill protrude from the fluff, a toy Miss Havisham in her white mink bed-jacket, warily watching for your intentions.

The system, at least when it works, is astonishingly effective. A chick that weighs less than 3 ounces at hatching comes to weigh over 2 pounds six weeks later, with some giant chicks coming close to 3 pounds, far more than any adult fulmar. They are obese, a third of their body made of the fats their parents have found for them in the squid, fish and plankton they have been catching and processing out in the Atlantic. They have loaded the next generation with fats – 2,600 calories of it by the time the chicks are six or seven weeks old, nearly three times the amount the parents carry. At this stage, the chicks look extraordinary, as inflated as a hot-water bottle, in that muddled stage where bits of fluff are still clinging to them, in among the first tail and wing feathers that are growing for their coming departure.

The making of these vastly fat chicks is an insurance policy, shared by the whole super-family of petrels, shearwaters and albatrosses. It may be that if the parents are unlucky in their hunting and can't return for many days, or if the wind doesn't blow and the fulmars cannot fly, the chick will survive only if equipped with that fat. Or, more likely, it is a preparation for those first hazardous days and weeks after the chick has left the nest and is relying on its innate hunting skills to survive. If it has some fat in the bank, a little parental deposit it can draw on before earning its living for itself, then its chances of survival will be higher, and the genes that animate both the young bird and its parents will have been served.

*   *   *

On one archipelago, 40 miles to the west of the Outer Hebrides, besieged by the Atlantic, where, as the sixteenth-century churchman Donald Monro warned, 'the streams of the sea are starke, and are verey eivil entring in any of the said iles', the fulmar was the sustenance of life. The islands of St Kilda are unlike anywhere else: deeply remote, with no easy landing, embedded in wild seas, which made fishing difficult, with their storm-swept cliffs visibly smashed by the Atlantic up to 200 feet from the surface on which you sail, and afloat on the astonishing riches of its seabird populations stacked in uncountable numbers layer after layer above you. For all these reasons St Kilda remained dependent on the birds long after they had become marginal for most Scottish coastal communities.

It was no paradise and the seabirds were not the food of choice. Seafood of any kind – and seabirds were thought to be seafood even by churchmen requiring fish for Lent – remained tainted as the resort of the desperate well into modern memory. There was an expression in Scottish Gaelic, *Beò air maorach a'chladaich*, 'living off the shellfish of the shore', used as shorthand for the pitiable condition of the landless. 'Down on his uppers' would make little sense in a world largely without shoes; 'forced to eat seafood' meant the same.

In late nineteenth-century Harris, Alexander Carmichael, the great collector of Gaelic thought and poetry, recorded a waulking song, sung as the women of the community thickened a baulk of cloth by beating its fibres, in which Mary MacLeod answered the accusations of a woman who was spreading rumours about her family and their poverty. 'Our homes are very different at sunset,' Mary spat at her enemy. 'In your father's house there are piles of fish bones; in ours we have mounds of venison haunches.' Bone analysis of the rubbish tips outside Hebridean houses from the Iron Age has shown that pigs were often fed on shellfish. Sea food, including the birds, was what you ate when you could get nothing else.

The St Kildans, with little meat and next to no fish, ate the birds, above all the fulmar, of which they consumed more than 100 each every year, plus the young gannets, the puffins and mountains of eggs. The prominence of seabirds in St Kilda life meant that from the seventeenth century onwards visitors recorded the seabird–human relationship in great detail.

Abundance was the raw material the islanders were working with. Every summer, the life of the ocean massed at the islands, raucous with its own vibrancy, a bellowing annual emergence of ocean DNA. For three or four precious months, the air over the colonies was as thick with life as a field of corn,

The fulmar harvest on St Kilda

a summer cavalcade and festival of creation, a million-bird gathering erupting each summer.

When, in the nineteenth century, outsiders first accurately measured the lives and habits of the St Kildans, they were killing 12,000 fulmars a year, taking the chicks just at the moment in early August of maximum oily fatness.

> During the preceding week an unusual excitement and alertness pervades the village. The women bring the cattle home from the sheilings, grind sufficient meal to last the killing time, while the men test the ropes, make good deficiencies, and provide barrels and salt ... The whole

village is astir and hard at work. A large and valued portion
of the winter's food must now be provided or you have to do
without it. The breeding places have all been examined some
time before, and an estimate made of the young birds which
they respectively contain. They are now divided into as many
portions as there are groups of four or five men who are to
work together ... The men either climb down to the
breeding places or get lowered by rope if necessary. The
birds must be caught by hand, or much of the valuable oil
will be lost. They must be caught suddenly and in such a way
as to prevent their being able to draw their wings forward or
they will squirt the oil ... Caught in the right way its neck is
speedily twisted and broken and the head passed under the
girdle [a rope tied around the fulmar-catcher's waist].

When the birds were brought to the top of the cliff, they were
given to the women and children who:

drained out the oil into receptacles, which are generally
made of the blown-out and dried stomachs of the Gannet.
This they do by the very simple means of holding the bird
downwards and gently pressing, when about a gill of oil [¼
pint] flows out by the bill ... When all are got home,
plucking off the feathers, disposing of the internal fat, and
salting the carcases for winter use goes on till far in the
night. Early the next morning the same round begins, and so
on from day to day till all the accessible breeding-places are
visited. All this time there is nothing but birds, fat, and
feathers everywhere. Their clothes are literally soaked in oil,
and everywhere inside and outside the houses nothing but
feathers; often it looks as if it were snowing.

The 12,000 oily chicks, handsomely provided for by their ocean-travelling parents, gave the St Kildans each year about 600 gallons of oil. Fulmar feathers were also collected and, more valuable than puffin or gannet feathers, they paid most of the rent up until about 1902. In 1847, each family had to give the landlord 7 St Kilda stone (1,176 pounds) of feathers, most but not all from fulmars. Feathers from eighty fulmars were needed to make a single stone.

Through the late nineteenth and early twentieth century, as James Fisher patiently discovered, the fulmar expanded from its ancient heartlands on St Kilda and in Iceland to colonize much of the North Atlantic. Fisher reckoned that the expansion was one beneficial effect of man's otherwise damaging impact on the ecology of the seas. The fulmars were feeding first off whale oil and blubber, gathering around the whaling ships of the north; and then off the discards and guts from the increasingly powerful industrialized trawler fleets. Man was shovelling food in the direction of birds which otherwise would have to hunt for themselves. The clarity of Fisher's story is in some doubt now – there is plenty of archaeological evidence from the ancient past of fulmars living and breeding in places into which Fisher thought they expanded only in the twentieth century. It may be that the expansion he witnessed was part of a regular, multi-century oscillation in which the birds came and went across the Atlantic, governed by long patterns of climatic change. But there is no certainty here: Atlantic fulmars are now in decline again, shrinking away from places in which they were thriving in the middle of the twentieth century. It may be that the reduction in North Atlantic fisheries, which have been ruined by overfishing, have again deprived the fulmars of the free food on which they had been relying.

But in ancient St Kilda the fulmar was more than just an economic crop. The islanders loved the fulmar, and the smell of the oil, which was 'a catholicon for diseases', an emetic, good for rheumatism, to anoint a wound or lessen a swelling, to help with toothache or boils, 'was sweet to his nose', as James Fisher wrote after talking to an ancient St Kildan fowler. He loved nothing more than to stir some dried fulmar flesh into his porridge.

We might like to imagine that our own connection to seabirds, thinking of them as a beautiful and entrancing part of nature, is somehow modern. It isn't. Birds have set up their territories in the branches of the human psyche for at least 60,000 years. And we have always had a double relationship to them: we have used them and loved them, nurtured and destroyed them, investigated them and sold them, looked at them with some awe and made toys, hats and dinner out of them.

That doubleness has evolved over time, but as the foundation layer, deep in our cultural consciousness, the root pattern is one of predation and reverence. Seabirds were never merely walking or flying larders but, as the Stone Age archaeologists Marcus Brittain and Nick Overton have called them, 'participants in life', fellow beings, co-actors in the drama and struggle of existence. People have consistently taken the birds and seen something magical in them. They have been food and poetry, metaphors for what we are or might be and sustenance for the poor. All over the world, in both hemispheres, seabirds have always been inhabitants of the great imaginative reservoir, emissaries from beyond the horizon, occasionally come to visit us in the mundane world, but just as ready to leave.

John MacCulloch, travel writer, geologist and social theorist, later author of *Proofs and Illustrations of the Attributes of God* (1837), arrived at St Kilda in the summer of 1819. His amazement was blanketed in a fog of condescension. 'If the High street of this city is a good deal encumbered with the heads, legs, and wings of birds,' he wrote of the old village on St Kilda, twenty-five or thirty ancient stone houses,

> these are probably entertaining enough to the inhabitants, reminding them of good dinners past and better to come ... The air is full of feathered animals, the sea is covered with them, the houses are ornamented by them, the ground is speckled with them like a flowery meadow in May. The town is paved with feathers, the very dunghills are made of feathers, the ploughed land seems as if it had been sown with the feathers, and the inhabitants look as if they had all been tarred and feathered, for their hair is full of feathers, and their clothes are covered with feathers. The women look like feathered Mercuries, for their shoes are made of gannet's skin; every thing smells of feathers, and the smell pursued us all over the islands.

His picture of a feather extravaganza is undoubtedly true: modern analysis of soils in St Kilda has shown that they are deeply contaminated with zinc and lead, absorbed by seabirds through the fish they caught, brought ashore with their bodies. The St Kildans manured their fields with birds and bird remains, mixed with the ash from peat and turf fires. But the manner of the amused, visiting gentleman, born in Guernsey, educated in Edinburgh, living in London, is symptomatic of a desire not to understand. There is, deliberately, an *Umwelt*-gap, no empathy.

Only a few hints remain of the St Kildans' own view of the birds, not as food, or simple protein to be preyed on, or part of the view, but as integrated elements in the landscape of existence. Travellers noticed that the youngest of boys would practise cliff-climbing on the stone walls inside their houses, waiting for the moment when they could be allowed out on to the bird cliffs. When the long winter ended, the St Kildans sang a song of euphoric welcome. 'Thanks to the Being, the gannets have come,' as Alexander Carmichael translated it in the 1890s,

> Yes! And the Great Auks along with them.
> Dark-haired girl! – a cow in the fold!
> Brown cow! Brown cow! My playful brown cow!
> Brown cow! My love! The milker of milk to thee!
> Ho ro! My fair-skinned-girl – a cow in the fold!
> And the birds are coming! – I can hear their music!

This is a form of poetry that has its roots deep in the ancient world. There is no straightforward moral instruction in it. The song relies on the power of suggestion, on the knowledge that a hint goes deeper than something said. Nor does it describe the elements it sets in play, merely placing them next to each other, much as they are in reality, allowing the listeners or the singers to establish in their own minds the connections between them.

So what does it mean, this St Kilda song about the coming of the birds? That their arrival in the spring is a miracle of fertility; that the miracle is dependent on 'the Being', the God of creation; that their renewed presence is as rich and delicious, as fertile and fruitful as new grass, the milkiness of cows, their playfulness in spring, and the expanding eroticism of the opening year. It is a form of celebration profoundly unlike our analytic understanding. For the song, things make sense when seen as part of a whole; different elements of the world are mutually illuminating; seabirds are happiness are spring are food are a well cow are milk are well-being and wisdom.

Eating birds, revering birds, becoming birds: the preservation of an oral culture in the Hebrides, and the necessary intimacy with the cycles of the natural year, the fulmar's own timetable geared to the light, warmth and emerging fertility of the year, kept a window open until the twentieth century into an earlier more connected world. For traditional societies on coastlines all over the world from the earliest times until now, seabirds and their eggs have been more than just an entertainment or a source of wonder. They were the source of protein, of fats and feathers, of clothing and insulation, of fuel for their lamps. In some places, as in St Kilda, seabirds penetrated every corner of life. Their pillows and mattresses were stuffed with them. Their houses were infused with them. Their diet depended on them. Their gardens were manured with them. But that physicality was also translated into something more. For the St Kildans – as for the Faeroese, the men of Rathlin and of the Blaskets in Ireland, the men of Grímsey and the Westman Islands in Iceland – the sense of their own dignity and manhood relied at least in part on their dazzling and courageous abilities to gather this harvest from the wild. They were in many ways people of the birds, seeing them as more than food and fuel, as part of who they were. Men might woo them on the cliffs; the women might sing bird songs over the boiling pot or the open grave. The smell filled their lives. It was the smell of home, the smell of who they were.

You can hear this ancient relationship now online. In the summer of 1951 the great American ethnographer Alan Lomax – the first man to use a tape-recorder to collect traditional music – went to Barra in the Outer Hebrides. There, in the house of Mary Gillies, he recorded songs sung by her friends Annie Johnston and Rachael Macleod, the two of them giggling at Lomax's mispronunciation of Gaelic. They begin at his urging, but the making of tea is going on in the background and Lomax has to break in: 'Hold the tea for a moment, Mary, please, because I am hearing every little rattle of the dish.' So the two ladies start again:

Iteagan, iteagan, uighean, Feathers, feathers, eggs,
Iteagan, iteagan, eòin, Feathers, feathers, birds,
Iteagan, iteagan, uighean, Feathers, feathers, eggs,
O 's e mo nighean a nì 'n ceòl. Oh it's my daughter can make
  the music.

Then the chorus, without words, merely bird sounds, which Rachael says 'is just diddledum. It's the seabirds making the music,' impossible to transcribe, the heart of what they meant to say.

The richest of fulmar myths has survived among the people of the Arctic. From Greenland to the Canadian north-west territories, Sedna, the terrifying goddess of the sea, is known also as mistress of the underworld. She is there only because of what happened to her on earth, and her ruinous tangling with the power of the fulmars. As a young girl, she had lived with her father in a remote part of Greenland. Her mother was dead and Sedna grew up to be a beautiful woman. People heard of her great beauty and from all over that part of the coast the young men came in their kayaks to woo her. She would have none of them until, one springtime, as the ice was breaking, and the birds were arriving with the returning light, a fulmar flew from over the sea and started to speak to her. 'Come to me,' he said, 'come into the land of the birds, where there is never any hunger. My tent is made of beautiful bird skins; my brothers, the fulmars, will bring you all your heart may desire; their feathers will clothe you; your lamp will always be filled with oil, your pot with meat.'

Sedna could not resist him and they went together to the land of the birds. It was a long journey over the sea and, when at last they arrived, Sedna found that her fulmar had lied to her. The tent he showed her was covered with old fish-skins, full of holes through which the wind blew and snow filled the tent. Instead of beautiful white caribou-skins, her bed was made of hard walrus hides. And she had to live on the miserable fish that the birds brought her.

She had thrown her life away, too proud to accept the young Eskimo men who had wanted her. In her grief she sang:

> Aya! Father, if you knew how unhappy I was, you would
> come to me, and we would hurry away in your kayak over
> the waters. The fulmars treat me unkindly, a stranger to
> them. The cold winds blow around my bed. They give me
> miserable food. Come and take me home!

After a summer and winter had passed, and the warmer winds once again started to blow across the sea, her father went to visit her. She fell on him and begged him to take her home. The fulmars were out hunting and so he quickly put her in the kayak and they set off. When the fulmar came home and found Sedna gone, rage overtook him. He called on the other fulmars and they flew out across the sea in search of the runaway girl and her father. Soon, of course, they found them and stirred up a powerful storm against them. The sea rose in enormous waves and threatened to drown the human beings. In fear of death, the father decided to offer up Sedna to the fulmars she had betrayed and threw her overboard. She hung on with a death-grip to the side of the kayak and her father took a knife and cut off the first joints of her fingers. Falling into the sea, they became seals. She still hung on to the kayak, more tightly than before, and her father cut again, the second joints of her fingers, and they too swam away as seals. When the father cut off the stumps of her fingers, they became the whales.

The fulmars thought Sedna had drowned and they allowed the storm to subside. Her father allowed her to climb back into the kayak. But she hated him. When they were ashore, and he slept, she told her dogs to gnaw off his hands and feet. Waking, he cursed her, the dogs and himself, and the earth swallowed them all. That is why Sedna now reigns in the underworld.

It would be difficult to find a more haunting seabird story. Everything about the fulmar – his seduction, his allure, his wide-ranging mastery of the ocean, the promise of riches, the actual desperation of his life, his oiliness and meatiness, his beautiful feathers, his ferocity and rage, his power, his intimacy with the ways of storm and sea, his dark, perceptive eyes and above all his mystery, his inhabiting a world beyond the one we know – is folded up into the myth. All that modern science is coming to know of the bird is somehow pre-embedded in this story, setting the fulmar in the place he belongs, as the core-spirit of the northern ocean world.

In the telegraphic language of the tale as it was told in the 1880s to the pioneer anthropologist Frank Boas, the fulmar is a shape-shifter, able to seek out Sedna and her father in the ocean through his 'magically transformed all-seeing eyes', *artsorutika akuleka taimaitjut ijingit*. When he does find them and summons the storm that he hopes will destroy them, he makes his triumphant cry, '*ia ha ha ha ha!*', the long and bitter laugh you will still hear from every fulmar in the world today.

2

Puffin

The year is late, it is almost May and in the distance the snow is still lying on the hills in Skye and Torridon. The grass on the islands is as dead as if it were the middle of winter, as pale and bleached as a hillside in Iran. Primroses are just coming out in the warm corners on the clifftops and a single red campion is in flower beside them. Razorbills are ashore and repeatedly, happily, contentedly mating on the rocks. The shags have already laid their long, dinosaur-oval eggs and are sitting over them, wearing their tall, party-ish, here-to-breed quiffs, the 'shags' that distinguish them from cormorants and give them their name.

But there are no puffins and without them the place is scarcely awake. All morning, hail has been falling in needle showers. Here above the islands, the skylarks are pouring out their song and the snipe are jumping in the marsh, a skitter and bang of life from the reeds. The shags might be all gloss and gleam as if made of freshly washed glass; the razorbills after coupling might be sitting side by side, nuzzling and billing with their old mate; but no puffins is no spring.

I wait and, one evening, the river of life comes streaming in out of the North Atlantic. The puffins have been away since August and now they return from nowhere, a sourceless gift

from distant skies, hundreds of thousands of birds, as if some
kind of life-nozzle had been swivelled in this direction. They
swirl and eddy towards the islands, the confetti of existence.

It is not often that you get a chance to witness the mechanisms of the world turning in front of you, but this is one. No
summer egg can be laid on the surface of the Atlantic. The
birds must return to these islands and so these evenings on the
margin of winter and spring are a kind of necessity. If the birds
are to exist at all, this flood tide has to come in. It is a fair-
ground of brilliance, high above you, lit by the last of the sun,
driven by sex. Sex is why every cubic foot of the island air now
flickers with bird bodies, alert with life, urgent with demand.

Watching the puffin wheel on the Shiants

The spectacle shares something of the Feeding of the Five
Thousand: so much from so little, so fertile in a place of stone
and salt sea, a flush of existence like a desert in which rains
have summoned the flowers. This arrival, when the birds glow
as seeds against the evening, is an ocean giving birth.

Deep time has played its part in shaping this pattern of absence and presence. The seabirds are the ancient inhabitants of these places and have been here longer than any of us. The oldest seabirds, shag-like with feathered wings and webbed divers' feet, are 100 million years old. We are the latecomers, African creatures that left our native habitat 50,000 or 60,000 years ago, the clever dominators of the earth who have learned to be where our bodies do not belong. If we are the temporary colonizers, the seabirds are part of the world on which we prey and feast. The average longevity of a mammalian species is no more than half a million years. We will be gone tomorrow. Many of these birds are older than the rocks on which they stand.

Even before the extinction of the dinosaurs 66 million years ago, birds had expanded into all kinds of vacant niches. Many of the modern lineages had existed before that catastrophe changed the shape of the world. In the next few tens of millions of years, birds looking like ostriches, owls, quails and all our tree-perching songbirds emerged. Most of them would have been strange to us: parrot-like birds without parrot-like faces, flightless owls and hawks stalking their prey like leopards in the undergrowth, and enormous pseudo-toothed seabirds, the biggest ever to have flown, which dominated the world's oceans until 3 million years ago. The true giant, the 30-million-year-old *Pelagornis sandersi* (a Dr Albert Sanders dug it out of a sandstone block under a new terminal at Charleston Airport, South Carolina in 1983), had a wingspan of 24 feet, double the biggest albatross, each of its primary wing feathers nearly 3 feet long. The bird weighed 8 stone, can only have launched itself off cliffs and would, like the albatross, have spent most of its life cruising the oceans, picking fish and squid from the surface in its giant serrated jaws.

The birds' lives and bodies are a response to the ocean. This evolutionary story is deeply dependent on ocean currents and

they also have their history. The great change in the making of
the North Atlantic came between 3 million and 2.5 million
years ago. Until then the sea sound between North and South
America, known to geologists as the Central American Seaway,
was still open and a current of relatively cool and fresh Pacific
water flowed through it into the Caribbean. The puffins, whose
ancestor lived in the Pacific, first arrived in the North Atlantic
5 million years ago, coming from the Pacific through that
ocean gap before it closed. Salty Caribbean water had long
been diluted by that Pacific stream, but when the land rose in
Panama, the Caribbean was cut off and it became the source
for very warm and very salty water driving north in the
Atlantic – the birth of the Gulf Stream, which has been the
governing giant in our lives ever since, flowing north at 3 cubic
miles of seawater an hour, 60 miles wide, up to 4,000 feet
deep, a tongue from the tropics licking up to high latitudes.

That vast stream had a counter-intuitive effect: as its warm
waters push north, enormous amounts of moisture evaporate
from the surface, fuelling northern rains and feeding the rivers
that drain into the Arctic Ocean. Once out in the sea, that
fresh water freezes as sea ice. The Gulf Stream is, in effect, a
snow gun, hosing moisture into the cold. Those frozen, fresh-
ened oceans turned the Arctic white and 2.74 million years
ago led to the most important event in the history of Europe
and North America: a climate crash, the sudden beginning to
the era of Ice Ages in which we are still living.

Since then, the North Atlantic has been either gripped in an
Ice Age or recovering from one. The variation between the two
states is dependent on a wobble in the earth's rotation every
40,000 years or so. When the axis of the earth tilts the northern hemisphere further away from the sun, cooler summers
allow ice to remain frozen all year. Each year the ice load
thickens and the north whitens. That whiteness reflects still

more of the sun's heat and the north is gripped by cold. As the axis drifts back, warmer summers melt the ice and the north returns to life.

The North Atlantic is far more susceptible to year-long ice-lock than the Pacific because it has a wide connection to the Arctic Ocean between Greenland and Europe. The Pacific's route to the north has always been either narrow through the Bering Strait or often non-existent when so much water has been tied up in the great continental ice-sheets that the sea level drops – as much as 280 feet – and a Bering land-bridge appears. The North Atlantic is also the smallest of the world's oceans, with a correspondingly smaller heat budget, more susceptible to temperature fluctuations in the cold landmasses around it than the Pacific to the west. The Atlantic, in these enormous terms, has been repeatedly subject to the devastation of ice. It has never known a long run without fundamental change. It is, at root, the ocean of change and trouble; the Pacific, well named, the sea of constancy and planetary calm.

The widths of the Pacific continued unaltered for millions of years. Temperatures scarcely dropped there in the Ice Ages. Generation after generation of Pacific birds were able to evolve in an almost completely stable world. Birds which somehow or other had arrived on remote islands branched into different species. In the Atlantic, there was hardly time to do that between the Ice Ages. There are relatively few tropical islands. Some, like Cuba, did follow a quasi-Pacific pattern, with large flightless owls and hawks stalking their pristine jungles, but elsewhere in the Atlantic endemics – species confined to particular places – only rarely evolved.

What you see when the puffins arrive in the spring is a product of this history. The Atlantic for the past 2.74 million years has been a place of coming and going, unsettled at the deepest of levels, a system always ready to flip from relatively beneficent to deeply unaccommodating. Life does not have the time here to develop the mass of differentiated variety it has within the security of the Pacific. Even the insects of the Atlantic have had to come and go, blowing back in with the winds when the climate allows them to survive, driving them out when it turns for the bad.

The result is that now in the North Atlantic there is relatively little local variation. Species have evolved to cope with the variability and have wide ranges across the latitudes. The Pacific is a mosaic of local land-based varieties; the Atlantic the exclusive realm of the ocean travellers, birds which have distance embedded in their ways of being. The ocean has been stirred into a single soup by the climatic traumas to which the ice has repeatedly subjected it.

Even now, when I am sitting in a bird colony in the North Atlantic, the multi-million-year history of this ocean and its climate shapes what I see and feel. Before the crash of 2.74 million years ago, before the Gulf Stream started to flow and the snow-gun blew across the ice-fields of the north, before ice began to hang over the Atlantic as an ever present threat, this was as multiple and speciated a place as the Pacific. Before the crash, the bony-toothed pseudodontorns would have been flying here. There were five species of albatross in the North Atlantic, off the coast of Suffolk, cruising North Carolina and Florida, at home in Belgium, probably crossing from Pacific to Atlantic and back through the sea channel at Panama before it closed. There is none now, except the occasional wanderer, lost in the wrong hemisphere.

Extinct species of shearwater, gannets and skua flew, fished and bred here. There are only three sorts of cormorants in the Atlantic now but there are six in the Pacific. Four species of auk are found in both the Atlantic and the Pacific, two are exclusively inhabitants of the Atlantic, but seventeen different auks can now be found in the Pacific. In the pre-ice Atlantic, this ocean would have matched them, knowable now only through the names the palaeontologists have given them: *Alca antiqua*, *Alca ausonia* and *Alca stewarti* were all relatives of the razorbill. The great auk was here, as was its cousin, *Pinguinus alfrednewtonii*.

There were as many kinds of puffin here as there are now in the Pacific. On these rocks, I would have seen birds not unlike the horned puffin, its blanked white face decorated with florid Mikado-style eye ornaments, versions of tufted puffins with their long, swept-back, film-star head-feathers and relatives of the rhinoceros auklet, dark, modest but with a proboscis above its beak which gives the bird its name. All were here, nesting and breeding on Atlantic islands, filling the summer air above the cliffs and skerries of the Atlantic, but all except the Atlantic puffin are now gone, erased by the glaciations that destroyed their feeding grounds, sank the ocean 250 feet or more and pulled the shore miles back from the colonies where they had bred. We are living in an ice-diminished world, one which the Atlantic puffin has made its own.

The birds have been preparing for this moment. They have been getting ready for summer, and since the experiments by the French endocrinologist Jacques Benoit in the 1930s, scientists have realized that the increased quantities of light falling on them has for a few weeks been stimulating a cascade of effects in their bodies and minds. For all birds in temperate latitudes, whose breeding cycles need to be attuned to the seasons, the abundance of food in summer means chicks will survive if born and raised then, and that simple invitation has shaped their life.

Scientists have been cruel to birds in pursuit of this knowledge: they have blinded them and understood that light does not reach the bird brain through the eyes alone. The pineal gland, which has its roots deep in the hypothalamus but emerges on to the upper surface of the brain, is itself light sensitive; blinded birds can recognize the days of summer as well as sighted ones. In very dim light, in which those blinded birds did not respond, scientists have plucked the feathers from the birds' heads and found then that the light sank into the brain through the translucent bone of the skull. When they injected Indian ink into the upper brain, the blinded birds again did not know whether it was light or dark. The pineal gland turned out to have cells in its surface very like the photoreceptive cells in eyes. Nocturnal birds, owls and the night-flying shearwaters and petrels, have very small pineal glands, but all the others are, quite literally, seeing sunshine with their brains. Even when scientists have implanted orange-red radio-luminescent paint in the brains of birds that were blinded and kept in the dark, they have imagined that summer has come again. The response is not only in the pineal gland. In several parts of the brain, light-sensitive receptors can bring about the changes needed for a successful summer; Benoit himself pushed a rod of light-transmitting quartz deep into the hypothalamus of a blinded duck and shone light down it, and by that disturbing method established that even there, at the very deepest levels of duck consciousness, increased light was understood as the trigger for an almost total transformation of the organism.

The changes these puffins have undergone over the last few weeks are sexual and social. Light is releasing hormones which are starting to convert the winter birds into their summer selves. Spring is happening in their physiology, in their minds and their desires. In the males, the testicles, which have shrunk over winter to the size of poppy seeds, are starting to swell. Soon they will be the size of broad beans. In the females, the oviducts, by which the egg is delivered to the outside world, is moving from its threadlike winter state to the massive muscled tube through which the bird will lay an egg that is not far short of 20 per cent of its own body-weight, the equivalent of an 11-stone woman giving birth to a 30-pound child. In the birds' brains, those parts that are responsive to visual signalling and are capable of song, which, over winter, have shrunk away, begin to grow back towards the size and condition they were in last summer. Winter puffins, dressed in grey, float in silence, picking at fish and plankton alone on the surface of the sea. The birds have started to eat more and put on weight to prepare for the demands of laying an egg and raising a chick. After moulting for most of the winter, they have now acquired new flight feathers with which to fly repeatedly from burrow to feeding ground and back, and a thick mat of warming breast feathers with which to incubate the egg.

Newly equipped for the summer

These deeply readied birds are the beautiful, perfected creatures you see gathered in spring at the colony. There is no sign in them of the rigours and challenges of a winter out on the Atlantic. They arrive and seem, as in many ways they are, newly made, about 7 inches high, weighing about a pound, little packages of exactness, their bodies neat billboards for their own health, their suitability as partners.

When, in Iceland, hunters go out to the summer cliffs with their fleyg nets to catch a puffin supper, they paint the rocks orange by which they will sit, put orange flags in the turf beside them and like to wear bright-orange clothes. At some time – no one can remember when – they discovered that orange drew in the puffins. Only in the last few years have bird scientists thought why. It seems as if the Icelandic hunters, by trial and error, learned to exploit a signalling mechanism that has evolved in puffins to demonstrate to each other that they are the sort of healthy, productive bird that any other puffin would want to mate with. As the breeding season comes on, the rather dark and dull-looking winter puffin – both sexes the same – is transformed by the hormonal tides surging within them into a breeding showpiece. Their beaks, feet and legs all turn a variety of a deep and vibrant orange, visible at a distance, standing out sharply from the black, white and grey of their plumage. A little orange rosette develops at the corner of their mouths, and the lower edges of the eyes acquire a bright-orange rim. They colour up in exactly the way the Icelanders dress to attract them.

Why should orange be so alluring to other puffins? The 'handicap principle', first proposed by the Israeli biologist Amotz Zahavi in the 1970s, is the key. Zahavi suggested that for any bird or animal signal to be effective, it has to be costly. If any creature could make a particular sign – a big tail, clever flying, giant horns – it wouldn't be worth making. Only if the signal cost a lot, in terms of body-energy or the difficulty of keeping it going, would it show that the animal making it was top quality. Expensive signals are the confirmation that a bird is good to breed with.

A very orange bill and rosette and very orange feet and legs are expensive for the puffin as the colour comes from concentrating large amounts of the pigments called carotenoids which are widespread in fish. Only if you catch lots of carotenoid-rich fish will your bill and legs turn a satisfactory colour. But the carotenoids are more than just a colour; they are also antioxidants, regulating the metabolism and stimulating the immune system of the fish that contain them – and of the birds that consume them. Puffins with lots of carotenoids in them are going to be healthier and better breeders than those which haven't managed to concentrate the colour.

The face as display-board

The puffin body is a placard for its own health. They are tentative at first, apparently reluctant to come back on land, now and then just dipping a toe in that strange world, anxious and wary when they are there, quickly returning to the sea. They gather in wide rafts, and there they mate on the surface of the water, most of them with birds they have mated with before: the annual divorce rate among British and Canadian puffins runs between 7 and 13 per cent; among Norwegian puffins, one study found the annual adultery rate to be precisely zero – and the death rate of spouses most years is less than 10 per cent. Look around you at a springtime puffinery and what you are seeing is largely familiar, paired-up sex.

Soon, they begin the most beautiful of their displays, the great wheels they establish over the colonies, in which tens of thousands of birds fly round and round, upwind over the colony itself, downwind a hundred yards or so offshore. The purpose is unknown; it may be that for many thousands of birds to fly together like this confuses the predators, the great black-backed gulls and the skuas, which could pick off individuals with ease. Or it may be a by-product of the aeronautical fact that landing upwind, when groundspeed can approach zero while the airspeed still allows the puffin to remain airborne, is the simplest and safest way of returning home, a policy all puffins try to follow.

Whatever the origins, the wheel is one of the marvels of the seabird world. In morning sun the birds are blown and lifted across the sky behind them, reflective chaff, like the metal glitter military jets emit to deceive pursuing missiles. Together they look like the circus of life, in which the wheel comes and goes, climaxes and fades, builds and intensifies, and then diminishes to a point where there is more air than birds, more wind than bodies carried on the wind. Puffins do not usually make a noise in the air as they fly, but in the wheel as they pass, at 15 or 20 knots, the air flutters and shuffles in their wings, the whispering of life taken at the flood.

If the wind changes or is fluky, the wheel turns in confusion. As it gusts, more than one wheel can develop, turning in opposite directions one above the other, or even one inside the other, or becoming for a moment a figure-of-eight, but never losing its animation. When a raven appears overhead, or a great black-backed gull on a cruise, there is a panicked dropping flight of the birds from their stances, a dump of black bodies, thickening the wheel like flour in a sauce. The puffins look out for and avoid each other, turning sharply in traffic, the wings folded to lose lift, or zigzagging through a blockage. Now and then there is a collision: I have seen a razorbill and a puffin smack into each other in midstream, a smart crack in the air, from which the two birds fell stunned and discomposed 150 feet towards the sea before recovering, shuddering themselves mid-flight into some kind of order and continuing on their way.

Watch them in a strengthening wind, so that they are flying into it almost at the same speed the wind is blowing them back, and they hang in front of you, 10 feet away, busy, looking resolutely forward and then sideways to see what you are, their features not sleek as they are at rest but ruffled, troubled, with the look of boats working in a tideway, more real and mysteriously more serious than the neat, brushed creatures you meet standing at the colony. That is a glimpse of the ocean bird, not on display but somehow private to itself, the bird that dies in winter, that in a bad year goes 300 miles or more in search of food for chicks it knows are hungry in the burrow, a bird at work, an animal whose life stands outside the cuteness in which we want to envelop it.

On land at the colony, the puffins are back in a place many of them have known for decades. They don't begin breeding until they are four, five or even seven years old, more often than not returning to where they were born, and after that coming back year after year. As adolescents, they explore the Atlantic, checking out other colonies, often flying around them, or standing on the edges in clubs, observing the adult breeders, learning the skills and geographies on which their successful adult lives will depend. The oldest puffin yet known is in its late thirties, but that is only a measure of how long people have been ringing them. It may be that puffins can live fifty years or more, pursuing the classic seabird strategy: long life, close, if not total, fidelity to the partner, a massive investment in one big egg per year, six weeks incubation, and then six weeks feeding the young chick.

They are accountants in the mathematics of life and their strategy is this: gold-plated funding of the new start-up, not spreading their resources too thinly among too many small ideas, all of which might fail, but banking on one new carefully set-up enterprise every year, understanding that the odds of surviving the competitive and environmental rigours of the Atlantic are low. They are not boom-and-busters but steady, long-term investors. They are destined to live in a high-risk environment. A variable food supply – will the fish be within reach of the breeding colony? – plus the prospect of a tough winter ahead for the just-fledged chick means that their tactics have to be building-society safe. No Byronic hell-raising risk-takers among the seabirds. The job is to live a long time, get yourself in good condition to raise a chick, find a good place to raise it, stick with the partner, don't start a family until you are old enough to be a good fish-hunter, and then every year make the most of your one nest-egg investment. Nearly all seabirds lay a big egg, raise that single chick, look after it with immense care, feed the chick up, so that some chicks end up bigger than their parents before fledging, and in some cases go on attending to it when it has left the nest and is already out at sea.

The birds know their neighbours exceptionally well, but behave with a kind of ramrod formality, so that the colony can look charming: Edwardian propriety, the sheer prettiness of their bodies, the butler-in-tailcoat walk, their own fascination with each other, rushing over to see what is happening if birds are billing, or starting on some sudden fluster of open-mouthed confrontation – but there is more than charm at work here. Colonies are huge clusters of very small territories, each belonging to a single puffin pair and each in competition with the others. In puffins, those territories, unlike those of their cousins the guillemots and razorbills, are sunk underground, in burrows a yard or so long, at the end of which the giant egg is laid and the chick raised. Dare to put your hand down a burrow early in the season and you will meet either a fierce, blood-drawing bite or magically, if the parents are away, the mysterious warmth of the egg, buried in the dark, resting in its shallow nest of grass, set up above the damp and slime of the burrow entrance, the shell at body-temperature on the tips of your fingers, an embryo in the earth-womb.

The burrow hides the egg and chick from gulls and rats, but just as likely it is there to protect them from other puffins. The colony is a world full of warnings and tip-offs, buried threats and hostile gestures, all conveyed with those demonstrative faces and bodies, as if this were an army of town criers, every puffin dressed in a near-identical herald's tabard, bright with maquillage and eyeliner, all saying Here I am, look at me, how good am I, watch what I am doing. On landing, each puffin bows and humbles himself, wings up, head down: no threat, no aggression. Then the bird shakes his head, shudders his wings into position, sorts out a stray feather or two and joins the group. The humility fades a little and with his lovely giant orange feet and legs, he walks around, either in an old-sailor, rolling gait or with comic largeness, up and down in his giant galoshes, as if walking on hot coals, sometimes not actually moving but walking on the spot like a cat playing with a mouse. All this occurs in almost total silence. Only occasionally does the bird make its slow, low groan-croak, like a soft and miniature chainsaw starting up, a little gunning of the throttle, a fading of the note at the end as the revs drop away.

The two sexes look the same and they acquire the coloured extensions of the beak only for the breeding season, when in the presence of other puffins. The circus outfit and face-epaulettes are dropped as soon as the bird heads out for its solitary winter at sea. For many centuries, they were thought to be different species.

It is tempting to say that the summer puffin beak is not a utilitarian structure. But that wouldn't be true, not only because it serves a function as an entirely effective prey-catching and nest-material-collecting instrument but because the social purposes of the beak are useful. It grows bigger with each successive summer, more deeply scored and ever brighter, until the moment the puffin is ready to claim or make a burrow and start to breed. You can tell a mature from a juvenile puffin by the glory of its beak. Breeding puffins, like hidalgos at court, need to look like heroes, an exhibition of their own marvellousness.

The historian David Crane has described how he saw the young Colonel Gaddafi at a contrived event in the Libyan desert in the 1980s, a figure exuding 'the bizarre mix of the charlatan and the real, of the charismatic and the stage-managed, of magnetism and latent brutality'. That looks like the world of the puffin, the arena of the pantomime kings. Gaddafi reminded David Crane of Napoleon's court, with its:

ridiculous titles, imperial flummery, bombastic military bulletins, state theft, court art, a controlled press, family greed, Corsican tribalism, ludicrous love letters, a sponsored cult, a lies and propaganda machine that might provide a template for any dictatorship – and all in the end for what?

All in the end for the display of power, as these birds are power players, efficient predators, tyrants of their trophic levels, massively successful, with more than 11 million of them in the world today, among fewer than a dozen seabird species that have reached that number.

Darwin in *The Expression of the Emotions in Man and Animals*, published in 1872, understood that the signals animals made were not just different from each other but fiercely contrasting. So that a dog that wants to be threatening, for example, raises its ears, tail and hackles and lowers its head; but a dog that wants to be friendly lowers its ears, tail and hackles, and raises its head. Exaggerated difference is the key because in a noisy and confusing world only exaggerated signals will make their mark. The puffin has to crouch and lower its head on arrival if it is not to be attacked; and then later walk tall with big feet to show its neighbours what it is made of.

The benefit of these signals is that they avoid the gull-like fights in which birds can get hurt or killed. Gulls which lay three or four eggs a season can afford the risk of a fight. For the long-lived, slow-breeding puffin, gull levels of violence would be a disaster, and to be avoided at all costs. Hence the atmosphere of theatre in a puffin colony: this is a place for show, for beautiful strong bodies, for perfect faces, for displaying the facts of their own excellence. Occasionally, there is a fight and a tussle between birds in which they go rolling down the hillside, bill clamped to bill, but on the whole they avoid confrontation, standing alone, held in a silent net of body-signals.

Why, if other puffins constitute such a threat, gather in a colony at all? Surely life would be better alone, hidden in some solitary nook or crevice like the black guillemots? There are all sorts of answers, none proven and none dismissed. If birds live in colonies, it is thought, predators will be quickly satiated with prey and many more chicks will survive than if their parents tried to live alone. It is the protection of the shoal. The colony is also a meeting place where birds can find a mate. And, maybe, through the example of other birds around it, the sheer communicative noise of a mass seabird gathering, the colony stimulates a bird to mate and lay. These advantages need to be set against the threat represented by the tens of thousands of rivals, all of which are after the same food, the same nesting places and the same top-quality mates as the others. A puffin colony, resembling in some ways a Canaletto-like gathering of sophisticated folk disposed around an elegant *campo*, is in fact taut between these imperatives. The burrows and the brilliant bodies are part of a single social system which is a form of frozen antagonism, where the mutual rivalry of the birds is held back by their withdrawing the precious chicks into burrows and, outside them, by engaging in dramatic I'm-not-going-to-hurt-you/Keep-Off signalling. What looks like calm is rigidi-fied stress and the sweetness is enmity stiffened into poise.

The work the adults must do at the colony is astonishing. The chick which weighs less than 2 ounces when it hatches has to be fed lots of fat, oily fish so that after a month it will reach the 12 ounces it needs to be if it is going to head out into the world properly equipped. If fish are thin on the ground and the parents have to fly hundreds of miles to find them, the costs become exceptionally high. Both parents drive themselves in pursuit of that goal, each diving between 600 and 1,150 times a day on the shoals of sandeels, sprats or capelin they need to catch. Most dives are fairly shallow, no more than 50 feet down, the fish probably visible in sunlight from the surface. But puffins have been recorded going much deeper than that, over 220 feet in dives lasting more than two minutes. When biologists measured the activity budgets of puffins from Skomer in West Wales, the average amount of time spent by a puffin under water every day during the breeding season was over seven hours, in addition to ninety minutes or so a day spent flying to and from the fishing grounds. They were working far harder, for example, than the razorbills, flying further, diving deeper and staying down longer.

The puffin wing must both propel it across the Atlantic and take the bird 200 feet or more down below the surface in pursuit of fish

It is a frantically busy time and both parents lose weight during the season looking after the chick, dragging in one food load after another, so that the chick is getting up to eight or even ten feeds a day, about 3 ounces of high-energy fish, in total worth about 100 calories. Each adult, to feed the chick and to survive itself, has to catch about 450 sandeels every day, at a metabolic cost to itself which has been calculated as the equivalent to a man spending his days knocking down brick walls with a sledgehammer. What you see at the colony, with the sea-washed puffins standing on the grass amused by their companions, are the brief moments of respite from a life of service.

This massive transfer of nutrients from ocean to island has always attracted man, and in the Blaskets, off the Atlantic coast of Ireland, the bird they loved best was the young of the puffin, right at the end of the summer after the parents had been feeding it up, the juiciest thing the sea could provide. It was called the *fuipín* or 'chicken of the sea', oozing with fat when roasted in a pot or grilled in a pair of tongs that were laid over the fire to allow the fat to bubble and run down into the glowing peats.

On the Shiants at the end of summer, young puffins wash about dead in the sea at the edge of the beach. If you pick one up, it is sodden like a small used towel, the same as a roll of loo-paper dropped by mistake in the bath. Even as your fingers close around the feathers on its body the water oozes from them. Press the grey face on the cheek below its closed eye and your fingernails are embedded in a mush of feather and seawater. With grey feet and claws, pre-sexual, no hint of display, but driven merely by the need to survive, the young puffin looks radically alone, dead at this most vulnerable moment of its life. Like the thirty others on this beach one cold summer day this August, it had not made it past that crucial horizon. Some had been eaten by the skuas and great black-backed gulls, others picked up by the peregrines and the sea eagles. Some had no marks on them. The Shiant black-backs have made a speciality out of eating puffin livers, puncturing a little hole in the abdomen, picking out the oily organ and leaving the rest untouched.

In August millions of other young puffins do make it away from their colonies. An extraordinarily high proportion of puffin chicks survive until they are five years old: 75 per cent in some Scottish colonies, even more in Maine and Norway. Both guillemot and razorbill fathers have long since taken their chicks out to sea, where they are now looking after them, but the puffin has another method. The chick or puffling stays in the burrow much longer than either of the other birds' young stay in their nests, and as a bigger bird, stronger and with more fat reserves, it then leaves to fight for itself, coming out of the burrow at night, unseen and unshepherded by the parents, which stay on at the colony for many days after the chick has left.

They too soon leave, and as they leave their physiology is going in reverse, back through the same processes they underwent during the spring. They must prepare for winter alone at sea. There is no need for the puffin to have all the breeding and social-signalling equipment which has dominated its life over the summer. Now the long July and August days trigger a massive decrease in the sex hormones. The ovaries and testicles of the birds start to shrink away. The egg-laying mechanism in the female shrivels. Their colour begins to fade, and once out at sea their beak and the patches on their face will all drop away. You can see it in birds even before they leave for the winter, a dulling of the spectacular display. There is no point in incurring the costs of being wildly orange when there is no more breeding to be done. Essentially, the birds are now returning to their condition before puberty, losing sexuality and losing adulthood. Unlike any mammal, but like many birds of the temperate latitudes, the puffins are going back to youth, without burden or demand, but with a focus instead on their own survival.

It is a dazzlingly effective system. About 95 per cent of all adult puffins return to the colony the next year, and until recently no one knew where they went or how they got there. There was a simple assumption that birds scattered into the ocean like peppercorns thrown on to a table, regathering only when the light of spring began to stir the desire for a mate in them. Recent tracking experiments by a team from Oxford led by the bird-migration specialist Tim Guilford have changed all that. On the island of Skomer off the Pembrokeshire coast, they tagged twenty-seven puffins with geolocators, which by measuring the times each day of dawn and dusk would record where the puffins had been over the winter. The geolocators were retrieved the following spring, and some birds were refitted with the trackers for a subsequent winter, so that between 2007 and 2010 Guilford's team gathered the puffin movements over forty-nine bird years. The results they came back with were revolutionary. Each puffin followed a different route from other puffins. They were not travelling around the Atlantic in gangs. That was to be expected, as 81 per cent of all winter sightings of puffins in the past have seen them alone on the ocean. But the revelation was that each individual puffin repeated a complex pattern of movements, unique to itself, winter after winter.

Every year puffin EJ09593 leaves the isthmus on Skomer in August and heads straight off to the ocean south of Greenland. It spends September there and in October moves east towards Iceland. A month later it is heading south and by Christmas it is off the west coast of Ireland, staying there until February

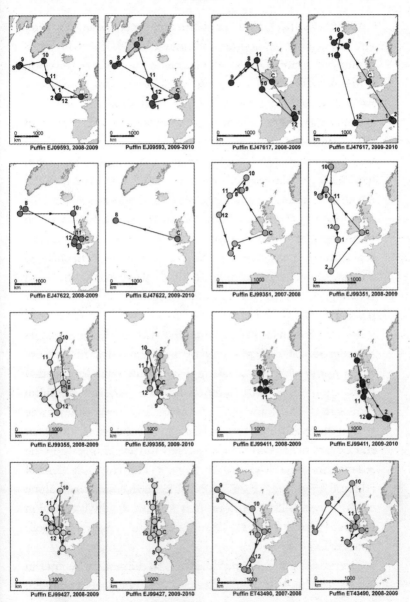

Individual puffins make highly repetitive winter journeys, often similar
to the journeys they had made the winter before but unlike those made
by other puffins

when it heads back to Skomer. The same route year after year. His next-door neighbour in Pembrokeshire, EJ47617, does something completely different, heading out into the Atlantic south of Iceland for the autumn, staying there, sometimes nearer, sometimes further from Iceland, before heading far south in November and spending Christmas and the New Year in the central Mediterranean. The coarse filter of the geolocators shows that puffin flying overland across France and Spain but that is certainly wrong; his route will be in and out of the Strait of Gibraltar. Others restrict their movements to the Irish Sea, the Minches and the Western Approaches, or track back and forth to the hotspots on the Mid-Atlantic Ridge. The repetitions are not exact, weather will drive them to and fro, but if you ask where a puffin is likely to go in the different phases of the winter, the surest guide is where it went the year before.

Previous theories of migration have assumed that birds are either genetically programmed or taught, and that young birds either know where to go innately or follow their parents and learn from them. This puffin behaviour does not fit either assumption. The young birds leave without their parents and so can learn nothing from them. And these individualized tracks are, at the very least, extraordinarily unlikely to be programmed for by the genes. Instead, Tim Guilford hypothesizes that the young puffins head out into the ocean on initially random tracks. They explore their Atlantic world, many finding the riches of the Mid-Atlantic Ridge, others somehow flying a long way south and east – ringed puffins have been recovered from beaches in Morocco and the eastern end of the Mediterranean – learning the ocean as they go. Some don't venture far; others embrace millions of square miles of sea.

Those that find the richer sources of food will survive; those that don't will never breed. Having made their initial explorations and discoveries as young birds, that information some-how becomes fixed in the puffin's mind. Each is now the bird that, for example, splits his winter between Iceland and Majorca; or flies between the Greenland Sea and the Irish coast; or stays near Pembrokeshire; or occasionally ventures an expedition to the Hebrides. They learn the land- and sea-marks that guide them between their various winter resorts. They are individuals, leading astonishingly individualized lives, swinging between their adult, sexualized, colony-located summers and the winter return to their pre-pubescent bodies. Fascinatingly and suggestively, when they become physiologically sub-adults, they again follow the tracks they discovered when they were young. The puffin body, in its different seasonal phases, seems to encode two different ways of being, on both of which the survival of the species depends.

Next time you sit among the puffins on a summer evening, looking at their elegance and anxiety, that is what to hold in mind: not clowns but beauties, Ice Age survivors, scholar-gypsies of the Atlantic, their minds on an everlasting swing between island and sea, burrow and voyage, parent and child, the oscillating nomad-masters of an unpacific ocean.

3

Kittiwake

I have once, after a difficult and anxious journey in a small open boat across the Minch, had a kittiwake come and hang over me, a few feet above the top of the mast, looking down and inspecting me and my belongings in the open hull as if I were the strange creature and she the scientist or the proprietor of this sea world. I had been worried for hours and the muscles had tightened across my chest in the difficult tide-cut waters. One or two seas had come aboard the open boat and I had been pumping and bailing, but now at last I was approaching safety and an anchorage, away from the wind, with a stilling sea, and suddenly this bird, a companion, even a form of consolation, floated above me. I felt then that I had never before been in the presence of such a sprung and beautiful thing, dawn grey, black eyes, black tips to the wings, the body held there as if on wires above me, afloat, dancing, its whole being like a singer's held note, not flickering or rag-like, nor blown about like a tern, but elastic, vibrant, investigative, delicate, the suggestion of a goddess momentarily present above me.

A seabird very like that appears in the *Odyssey*, the first sea poem of our culture, certainly 3,000 years old and with elements up to a thousand years older. For Homer, seabirds

The kittiwake in flight

are bringers of salvation. When Odysseus is alone on his raft, sailing away from the clutches of Calypso, steering by the stars night after night, reaching for home, he is struck by a terrifying storm. The great man just about hangs on to life as trouble thrashes around him. Death seems near at hand until at last, mysteriously, a seabird, *aithuia* in Greek, maybe a tern, of which many species are seen in the Mediterranean, or a kittiwake – no one can be sure – alights on the wave-swept raft and speaks to him. Kittiwakes are regular winter visitors to the Mediterranean, sometimes found as far east as Cyprus and the coasts of Israel. They are not the commonest of Mediterranean birds, but frequent enough for them to have been known to the Bronze Age sailors that were pushing out into their inland sea.

Homer's view of the bird is the opposite of the Anglo-Saxon poets of 'The Wanderer' and 'The Seafarer'. In the northern poems, the sailor regards the seabirds as beautiful but inaccessible, part of the frozen world of loss in which his imagination likes to dwell. For the Greeks, they play the opposite role. The sea is a desert of hostility, stirred up by the rage of Poseidon who loathes Odysseus, the realm of death, threatening destruction at every turn, but out of that hostility the beauty of the *aithuia*, a sweet-eyed, perfectly feathered thing, comes and speaks to the man so helplessly exposed to the ocean's rage.

She is Leucothea, the white goddess, once a girl from Thebes, with, as Homer says, 'the most beautiful ankles', who has been taken up by the gods as one of theirs. She gives Odysseus a holy veil. He is to tie it around his chest, abandon his broken raft and swim for shore. What she gives him – and woven objects in Homer are always carriers of goodness – will save him from the chaos and sorrow of the sea. With those instructions, she leaves him, plunging into the swells, 'like an *aithuia*, and the dark wave hid her'. That is like a kittiwake: a picture of delicacy and tailored perfection, a sign of grace and goodness, feminine and intimate with the man, with you and then not with you, not hovering like terns over the prey, but diving straight in from direct flight, plunging up to 3 feet beneath the waves.

In part anyway, these are imagined qualities. This gentle-looking bird has conquered the world. No gull is more populous. Something like 18 million kittiwakes are spread across the northern hemisphere, their buoyancy not a girlishness but a powerful genetic fact. If you spend an evening above a kittiwake colony, you are looking at a seabird that has been, at least until recently, one of one of the more successful on earth, far more numerous than any of the bruiser herring or black-backed gulls that hang about our lives. And its great secret has been to leave that coastal half-zone in which most gulls subsist and make for the open ocean.

The colony will tell you easily enough that sweetness is not really their mode of being. Their greeting and welcoming calls echo against the solids of the rock, each calling as its mate arrives, each mate calling on arrival, *kittiwaaak*, *kittiwaaak*, a comb-vibrating harshness shot across the colony, the astonishing lipstick-red gape of their mouths and throats emerging suddenly in this world of grey and rock. One pair after another takes it up, and the noise ripples around the coastline in an aural swell breaking on the cliffs, *this is who we are, this is who we are*, as one of a pair leaves the nest for a moment, circling, before returning to its mate, just touching on the nest lip, before allowing itself to fall away again, 2 or 3 feet down, and then off in a sweeping clean-cut glide, its wings and tail spread and splayed for lift.

People have always listened to them trying out their name. The great seventeenth-century ornithologist John Ray called them Cattiwakes, Daniel Defoe heard them in Northumberland as Kittawaax, and in Newfoundland they are called Tickle-ass, said with a long Newfoundland 'a', another transliteration of the cry, and meaning 'the little bird that lives around a tickle' – the Newfoundland name for a narrow channel between islands. In Iceland they are called Rita, perhaps meaning 'to squeal'; or maybe after the girl's name, a measure of their sweetness and the darkness of their eyes.

The great change in understanding them, looking beyond the symbolism and prettiness to the bird's habits, came from one of Niko Tinbergen's students at Oxford in the early 1950s. At Tinbergen's suggestion, in the summer of 1952, Esther

Cullen, a young Swiss biologist, went to the Farne Islands off the Northumberland coast. Tinbergen, who was already conducting a broad campaign with several DPhil and postdoctoral students to understand the different species of gulls, visited her there and was excited by what he saw of the kittiwakes: 'A most intriguing species! The most aggressive of all I have seen; there is severe competition for nest sites, quite different from the situation in the Herring Gull.'

Cullen spent the long summers of the next three years watching the kittiwakes with the close attention to realities and the search for systematic patterns in body-form and behaviour that was the hallmark of the Tinbergen method. In a hide, wrapped in a sleeping bag, with hot-water bottles, sheepskin gloves, multiple jerseys and triple socks, she watched the birds for thousands of hours.

Drawing on those many frozen, limb-stiffening vigils, Esther Cullen described in her classic paper, published in 1957, the many characteristics of the kittiwakes she had been observing, sometimes alone, sometimes with Tinbergen and other students. She had developed a way of identifying individual birds by the slight differences in the dark markings on the tips of their primary feathers, and having identified them often gave them names.

One day in May 1954, she and Tinbergen sat together on the cliffs of Near Farne. Esther Cullen had named one of her sex-hungry birds Casanova, and this afternoon Tinbergen and she watched the results of ferocious competitiveness between neighbouring kittiwakes unfolding in front of them as if in the audience for an Italian opera. It could have been called *Measure for Measure, or The Cuckold's Revenge*.

Like all Italian operas, the plot, as described in Tinbergen's notebooks, needs some close attention. The first Act was a straightforward fight between neighbouring males, beginning when Casanova plunged off his own nest and attacked the kitti-wake next door. 'They fought by biting each other in neck, face and beak,' and the fight ended when Casanova threw the neighbour out of his nest. Casanova's wife had been standing on the ledge above the fray. As it reached its climax, he sang to her and she came down to him. At this stage, it was merely a land-grab. Casanova had conquered the neighbouring nest and made it his own.

Act Two began when the wife of the defeated and ejected male arrived back at the nest which Casanova had taken over. She immediately attacked him and then his wife. Confused fighting followed, with the two females jabbing at each other over Casanova's neck. For a moment or two, like the put-upon hero, he sat uneasily between them head down, but then, in the emotional turmoil of the moment, both women attacked him. 'For a long time they both squeezed his neck between their bills and he gaped several times "in despair".'

The real enmity was now between the two females. They set to and Casanova's own wife soon lost the fight with her rival. She was thrown off the nest, spiralling away into the abyss. Casanova and his neighbour's wife stayed there, 'as if in pair'. End of Act Two: 'The Adulterer's Masterpiece'.

Act Three began when the ejected male rival returned, flying up to the lip of what had been his own nest, where his adulterous wife was now ensconced with Casanova. The first time the defeated bird came up, he was not even allowed to land and was pushed away into the air. But he swung round, came back and this time landed. 'A wild struggle ensues which ends with Casanova being thrown over board. We all applaud: justice has been carried!' The re-established pair 'now indulge in tremendous display building' – the making, stamping down and preparing of the deep-pocketed mud-and-grass nest in which the kittiwake eggs are laid. Casanova had not, on this

Not much room on a cliffside nest

occasion, had time to mate with his neighbour's wife but, after all the fuss, it was clearly important for the would-be cuckold to reassert the idea that both nest and wife were his.

From the mass of data she gathered, Esther Cullen was able to establish that the way of life of a kittiwake, its habits and behaviours, could all be explained by understanding that it was a gull which, in evolutionary terms, had recently exchanged a shore-nesting, coastal life for nesting on cliffs and living on the open ocean. Like those of most shore-nesting gulls, kittiwake eggs have a very slight cryptic patterning to hide them from predators. But this is a hang-over: there is no need for camouflage because the kittiwakes have chosen to glue their nests to tiny rock-climber projections, spread across the cliff face like a village, always well down from the top of the cliff and unreachable by predators. Unlike other gulls, even as young birds, the kittiwakes have fiercely clamping prehensile claws with which to hold on to that nest in a wind or a gust. Those chicks are not camouflaged, as there is no need, and for the first thirty days or so, kittiwakes cannot distinguish between their chicks and others'. The capacity which other gulls have to know their children has withered in the kittiwake as their chicks cannot leave the nest and still survive. Nest material is rare on the cliffs, collecting it takes time and energy, and so kittiwakes have developed the habit of stealing material from other nests when they can. Nothing is more important than keeping the nest deep and strong, so the birds defend their nests ferociously. Nor do kittiwake chicks run away and hide when threatened, as other gulls do, but crouch, turn their back to the aggressor, hiding their bills so as to diminish any suggestion of threat or even defence. All of it suggested one cause to Esther Cullen: kittiwakes were reluctant returners to the land. The sea cliff was the nearest they would come to being land-dwellers again. These were all the adaptations of a bird that had adopted the deep ocean as its habitat.

She and other members of Tinbergen's team at Oxford wanted to know if these behaviours had become innate in the birds. Gull chicks of ground-nesting species tend to wander about; if kittiwake chicks did that, they would die. So did the chicks realize that leaving a cliff nest would be dangerous? Or was the tendency for kittiwake chicks to stay in the nest genetically programmed into them? Cullen and Tinbergen performed some fairly heartless experiments.

Esther and her husband Mike Cullen put some eggs of black-headed gulls and herring gulls, both ground-nesters, in the kittiwakes' nests. The kittiwake parents couldn't tell the difference, the eggs were accepted and duly hatched, but, as Tinbergen described it, 'most of the chicks did not survive very long, for, showing not the slightest inhibitions, they soon began to walk about with disastrous consequences ... There is, therefore, no doubt that we have to do with a real, innate difference between Kittiwakes and other gulls.'

Heather McLannahan, another member of the Tinbergen group, later took up many of Esther Cullen's suggestions and tested for the opposite question. What would a kittiwake chick do when exposed to the threat of falling over the cliff? Using a perspex sheet, she made an artificial cliff, in which it would seem to the chick as if the ground were falling away beneath it, even though it could stand quite safely on the perspex. She hatched some kittiwake chicks and kept them in an incubator. For the experiment, a chick was taken out, with its eyes carefully blindfolded, and placed exactly on the edge of the 'cliff', sometimes with its head over the abyss, sometimes pointing into the ledge.

When the blindfold was taken off and they were allowed to see, as Tinbergen later reported, the chicks panicked.

The results were very striking indeed, particularly in the first type of test. The chick would look round and would soon see the abyss, and then it would at once crouch, tremble all over its body, even spread its tiny wings, and try to hook its claws into the substrate. And at the same time it turned and crawled away from the 'edge'. A couple of centimetres or so from the 'edge' it would relax, the trembling would stop, it might even stand up and preen itself – all signs of being at ease once more.

Heather McLannahan's experiment with the perspex sheet and the false cliff

This was true of all kittiwake chicks, but not of herring gull chicks on which Dr McLannahan made the same experiment. Herring gulls, raised on level ground, do not need to know that a cliff will kill them. Only in young kittiwakes has a terror of dropping into oblivion evolved to save their lives.

The young kittiwakes are beautiful birds, their plumage quite different from the parents, with a black bar across the neck, and an M of black feathers across wing and back when in flight. They stay with the parents for many days after fledging and you can watch them trying out their newly mobile selves, dropping from the nest over the sea and then down, slightly uncertain, always turning at the same point, the same little headland at the edge of the bay, or an isolated rock in the sea, like a child on a bicycle practising his geographical skills, trying his route again and again, drifting here, veering there, turning there, a flutter within a quarter-inch of landing, but then the release away and down on to the track again, that dropping catenary route from the height of the cliff to a few feet above the sea like a skateboarder perfecting routines.

They were so beautiful that they became a favourite target for Victorian sportsmen. What was called 'shooting flying' – taking the bird on the wing and not having to wait until it settled – had been in vogue, if difficult, with the muzzle-loading flintlocks of the eighteenth century. Gradual improvements to guns – loading from the breech not the muzzle, double-barrelled weapons, the self-contained cartridge – gradually made shooting easier so that by the 1830s almost anyone could be out at the sea cliffs for a day or so. Colonel Peter Hawker, over 6 feet tall, strikingly handsome, 'up to the end of his life … very erect' and from a family that had been soldiering in the service of the English crown since the days of Elizabeth, loved nothing more than shooting seabirds but, to his chagrin, often found he was not alone in this pursuit.

Jan 13 1814. – The wild fowl at last came into the haven by thousands, in one continued succession of swarms, and in a few hours, notwithstanding this was a day appointed for a general thanksgiving, an immense levy en masse of shooters was assembled at all points, and there was not a neck of land, bank, or standing place of any kind but what was crowded with blackguards of every description, firing at all distances, and completely annihilating the brilliant prospects of sport.

The following day was, if anything, worse, with 'a numerous host of boats and canoes' filled with men out to get the birds.

> I had in all quarters the mortification to find myself closely surrounded by vagabonds of every description, who were standing quite exposed, firing at sea-gulls, ox-birds [dunlin], and even small birds, and repeatedly, as the geese were coming directly for me, like a pack of hounds full cry, I had to endure the provocation of seeing some dirty cabin boy spring up and drive them away with the paltry discharge of an old rusty popgun. Had it been possible for me to have lain peaceably in any one place, I should have filled a sack; as it was, however, I had no further satisfaction than that of killing more than all these ruffians put together. I got 3 wigeon, 2 grey plover, 2 cormorants, 1 ring dotterel, 18 ox-birds, and 1 dusky grebe.

All over the English coasts from Devon to Northumberland, people were killing the seabirds for fun. The numbers of kitti-wakes on St Abb's Head in Berwickshire were said by a local squire in 1794 to 'breed incredible numbers of young'.

> When the young are said to be ripe, but before they can fly, the gentlemen in the neighbourhood find excellent sport by going out in boats, and shooting great numbers of them; when they are killed or wounded they fall from the rocks into the sea, and the rowers haul them into their boats. Their eggs are pretty good, but their flesh is very bad; yet the poor people eat them.

When a regular east-coast passenger service was set up in the 1830s, the kittiwakes provided momentary diversion en route and were 'so much tormented by the large Newcastle steamboats calling in and firing at them' and 'so much shot at by the London and Leith steamers' that they deserted the headland. The same thing happened to the birds nesting on Gannet Rock off the north end of Lundy in the Bristol Channel. No Bristol cutter would pass by them without the pilots killing them for fun.

The profoundly eccentric Yorkshire squire George Waterton, a visionary and pioneering conservationist, recorded a visit to the teeming bird cliffs at Flamborough Head on the North Sea coast of Yorkshire in the 1830s. To his horror he witnessed:

> the unmerited persecution to which these harmless seafowl are exposed. Parties of sportsmen, from all quarters of the kingdom, visit Flamborough and its vicinity during the summer months and spread sad devastation all around them. No profit attends the carnage; the poor unfortunate birds serve merely as marks to aim at, and they are generally left where they fall ... [The birds] fall disabled on the wave, there to linger for hours, perhaps for days, in torture and in anguish ...

A few years later, a naval officer, Hugh Horatio Knocker, Commander in the Coastguard, reported that 'On a strip of coast eighteen miles long near Flamborough Head 107,250 birds were destroyed by pleasure parties in four months, 12,000 by men who shoot them for their feathers, and 79,500 young birds who died of starvation in the emptied nests.' Knocker 'saw two boats loaded above the gunwales with dead birds, and one party of eight guns killed 1,100 birds in a week'.

The delight in nature which swept through European and American culture in the eighteenth and nineteenth centuries took its toll on the birds. Feathers in the hair came to seem like the sign of a fine sensibility. The fashion ebbed and flowed but never quite went away. Marie Antoinette had begun it with ostrich feathers; ladies at Versailles developed such enormous towering feathery styles that they had to travel to balls kneeling on the floor of their carriages. The fashion lasted for decades. 'A well dressed woman nowadays', the *Los Angeles Times* told its readers in 1891, as if it were news, 'is as fluffy as a downy bird fresh from the nest.' 'If you would be fashionable this winter, you will be beplumed,' declared the *Chicago Daily Tribune* in 1910. When a man threw a 'keg of liquor' from the upper gallery at the Covent Garden Theatre in February 1776, only a lady's feathered headdress prevented it from crushing a member of the audience. One Saturday night in January 1796, Major Edward Sweetman, in the upper side-gallery of the opera house, found himself behind a lady 'whose head-dress of fashionable feathers prevented his view to any part of the house'. 'Leaning over the Lady's Head-Dress', Sweetman 'incommoded her', saying he couldn't see the play. Captain Brereton Watson, her companion, remonstrated with his fellow officer and, when the apology was inadequate, accused him of being no gentleman. A duel was inevitable. They met on the Monday morning on Cobham Heath in Surrey; the distance fixed was seven paces; both parties discharged their pistols at the same moment. Watson was shot through the right thigh, fracturing the bone, but a pistol ball went straight through Sweetman's chest. He did not speak another word and died before he was carried back to Cobham, where he was buried the following day, the only known human fatality of a bird-feather hair-do.

Young kittiwakes were in the front line. They were shot on the Farne Islands as early as 1802, and on the Isle of May, at the mouth of the Firth of Forth, as food. At Clovelly, in Devon, just across the Bristol Channel from the big seabird cliffs on Lundy, a cottage industry prepared plumes for much of the

The Extravaganza or the Mountain Head Dress of 1776

year. At the height of the season, 500 or even 700 birds were brought in for the pluckers in one day. One nineteenth-century estimate reckoned that 9,000 Lundy kittiwakes had been shot in a fortnight. That may be at the high end, but it is clear that the kittiwakes on Lundy (which is now down to a hundred breeding pairs and dropping) suffered a heavy and regular mortality.

For some ornithologists in the middle of the nineteenth century, the destruction of the kittiwakes became intolerable, not out of a feeling for the beauty of nature – that was what had turned them into women's hats in the first place – but out of a more science-based fear for the continuation of the species in Britain. Alfred Newton, professor of comparative anatomy at Cambridge, was also Vice-President of the Zoology and Botany Section of the British Association. At a meeting of the

Association held at Norwich in 1868, he read a paper called
'The Zoological Aspect of Game Laws', in which he did not
talk about 'bird murder' – the term used by most of the
anti-shooting lobby. He was simply interested, although this
word did not exist, in bio-diversity. The bustard and the large
copper butterfly had already gone. The birds of prey were
under threat. 'But now for Sea-fowl – and here I must plead
guilty to the charge (if it be a charge) of being open to a little
bit of sentiment.'

At the present time I believe there is no class of animals so
cruelly persecuted as the sea-fowl which throng to certain
portions of our coast in the breeding season. At other times
of the year they can take good care of themselves, as every
gunner on the coast knows; but in the breeding season, in
fulfilment of the high command to 'increase and multiply'
they cast off their suspicions and wary habits and come to
our shores … But how do we treat them? Excursion trains
run to convey the so-called 'sportsmen' of London and
Lancashire to the Isle of Wight and Flamborough Head,
where one of the amusements held out is the shooting of
these harmless birds. But it is not merely the bird that is shot
that perishes – difficult as it is to say where cruelty begins or
ends – that alone would not be cruelty in my opinion.

The bird that is shot is a parent – it has its young at home
waiting for the food it is bringing far away from the Dogger
Bank or the Chops of the Channel – we take advantage of its
most sacred instincts to waylay it, and in depriving the
parent of life, we doom the helpless offspring to the most
miserable of deaths, that by hunger. If this is not cruelty,
what is? … We thank God that we are not as Spaniards are,
who gloat over the brutalities of a bull-fight. Why, here in
dozens of places around our own coasts, we have annually an

amount of agony inflicted on thousands of our fellow-
creatures, to which the torture of a dozen horses and bulls
in a ring are as nothing.

He went on to report the opinions of a correspondent who
had recently outlined the kittiwake crisis at Flamborough on
the Yorkshire coast: 'One man, a recent arrival at Flamborough,
boasted to me that he had in one year killed, with his own gun,
four thousand of these gulls; and I was told that another of
these sea-fowl shooters had an order from a London house for
ten thousand.' It was no wonder the kittiwakes were rapidly
disappearing. 'Fair and innocent as the snowy plumes may
appear in a lady's hat,' Newton ended with a flourish, 'but I
must tell the wearer the truth – "She bears the murderer's
brand on her forehead."'

Wearing dead birds could demonstrate a love of nature

Newton argued that Parliament should legislate to create a closed season for the breeding seabirds, in which none could be shot. He had timed it right and the following year, in 1869, the Sea Birds Preservation Act was passed by Parliament, the first environmental legislation in Britain of which the main focus was not the preservation of a huntsman's quarry but the well-being of a wild creature. A special exception was made for the people of St Kilda, still dependent for survival on the annual seabird harvest, but in other ways the Act was far from perfect. It still allowed shooting of kittiwakes and other seabirds after 31 July each year, just when the young birds were first on the wing. As bird fashions in all parts of the developed world continued to gyrate through ever more baroque versions of the fad, many kittiwake populations continued to crash. In the United States, *Harper's Bazar*, as it was then called, faithfully reported the changes.

1875 Fall: Birds and small wings on imported French hats
1878 Summer: Heads and wings as wall hangings
1884 Winter: Feather fans, nodding plumes for younger girls
1884 Fall: dresses bordered with smooth soft feathers and birds' heads. All imported bonnets carry feathers for a 'wind-blown' effect.
1893 Fall: Valkyrie effect of Mercury wings *in extremis*; dinner gown of black duchesse bordered with tiny swallows with outspread wings.
1899 Summer: whole birds for walking hats; wings, feathers on golf-hats and horse-show costumes
1899 Fall: feather heaped windows (Seabirds and grebes).

In London it was no better; the problem had simply been exported to the Empire. In August 1885, a bird lover wrote to *The Times*, telling the governing classes that between December

1884 and April 1885 the London plumage market had sold
6,828 birds-of-paradise, 4,947 Impeyan pheasants, 404,464
West Indian birds and 356,389 East Indian birds of various
species. About 20,000 tons of ornamental plumage were
imported to the United Kingdom between 1870 and 1920,
when the trade was finally stopped, and the same quantity or
more was taken into France. Both figures represented 'scores
of millions of birds killed'. Newton's gesture for the kittiwake
was nothing but a beginning. Swallows may no longer decorate
the neckline of evening dresses and Newton's seriousness, not
as an animal lover but as a biologist, had at least set seabird
conservation on its first tottering steps, but even now, 150
years later, real conservation measures for the world population of birds are not yet in place.

*   *   *

Modern science has started to understand the genius of these
birds. It is easy enough to see in the dynamism and noise of a
kittiwake colony a kind of chaos, an anarchic grab at the possi-
bilities of survival, but some disturbing experiments at a
Norwegian kittiwake colony have revealed – as if this needed
to be revealed – that the kittiwakes do know what they are
doing. On the little seabird island of Hornøya, less than a mile
in each direction, high in the Arctic on the shores of the Barents
Sea, more than 90 per cent of the kittiwake nests contain only
two eggs. The biologists wanted to know what would happen
if the kittiwake couples were faced with raising either one
chick or three. For four summers, from 1990 to 1993, they
investigated the fate of sixty-seven kittiwake pairs on Hornøya,
enlarging twenty-two families from two chicks to three,
reducing twenty-nine from two chicks to one and leaving
sixteen uninterfered with.

Seabirds have to make a decision each year about breeding. They must navigate their way around a triangle of forces in which none of the corners is fixed: the number of offspring they are going to raise; the likely condition of those offspring; and their own condition after putting the effort in, with the implications that has for their survival to breed again next year. If they raise many chicks, all of which are strong, but the effort leaves the parents weak, that is no good for the future. If they raise many offspring, but attend too much to their own well-being while doing so, the chicks might be weak and that is not much good either. But if they think largely of themselves and raise very few chicks, and don't overstretch themselves, that is not much good for the future of the race either, however strong the chicks might be.

The Norwegian bird scientists started to interfere with the natural mechanisms of the Hornøya kittiwakes when the chicks were a week old. As no kittiwake recognizes its own chick for four weeks or more after they hatch, parents do not reject strangers. But the experiment, as faithfully reported in the pages of the scientific journal *Ecology*, soon entered the realms of disaster. Large numbers of the chicks died, either starving or eaten by gulls and ravens. Half of the nests that had been left with one chick lost that chick and every single one of the nests with three chicks lost one of them. In those nests that had been given three chicks, less than a fifth of the chicks survived to fledge, and those that did were thin. All adult seabirds lose weight while bringing up the young, but the burden of bringing up three chicks was making heavy inroads into the well-being of the mother kittiwakes, all of them weighing substantially less than the mothers of one chick at the end of the season.

The really devastating impact, though, came over the following winter. More than 90 per cent of the mothers with one chick survived to breed again the following year, but very nearly half of all the mothers that had been made to raise three chicks were not seen again at Hornøya the following spring. The scientists looked for them on all the neighbouring coasts and islands but could not find them and were forced to conclude that they had died over the winter. Or in their language: 'We suggest that the reduced return rate of females with enlarged brood size does represent a mortality cost of increased breeding effort.' The female kittiwakes, compelled to respond to the demands they couldn't meet, had pushed themselves to the point where they could not survive a winter. The males showed no such pattern, and continued blithely into the future, a gender difference for which the scientists could offer no explanation except to say that 'females respond more readily to changes in food requirements of the chicks than do the males'. This difference between male and female occurs in other species, but why that should be is yet to be explained. The one fact this experiment revealed was that kittiwakes already understood that from that particular island in that particular sea, their best option in balancing the demands of their own survival as individuals with the projection of their genes into the future was to lay two eggs and raise two chicks. We were not smart enough to guess how smart they were.

Their attention to their world turns out to be astonishingly detailed. David Irons, the veteran seabird watcher, put some radio trackers on kittiwakes from a colony in Prince William Sound, the forested and islanded body of water on the southern coast of Alaska. Big and turbulent tides, with a rise and fall of 20 feet twice a day, churn through the Sound, with upwellings and eddies where they make their way past the headlands and islands. These nutrient-rich waters are the places where the plankton concentrate and are magnets for seabirds, particularly in the two hours at the centre of the flood or ebb, when the tide is running at its fastest, just at the mid-point between high- and low-water marks, and particularly during spring tides when the volumes of moving water are higher.

Irons's tracking of the kittiwakes revealed an intriguing pattern. The birds did not go fishing at the same time every day but adjusted their schedule to coincide with daily tidal cycles. One of his radio-tracked birds used to fly every day, twice a day, 25 miles to the south from its nest at Shoup Bay to the tidal upwellings around Glacier Island and then back again. In the middle of the month, when the spring tides were running, it would be fishing from about 4 in the morning until about 7 and again from about 3 in the afternoon until about 6 in the evening, both times hitting the peak of the ebb. This was high-latitude summer and there was no limit on daylight. As the month went by, the tides moved from springs to neaps, with much lower energy in the system, and slipped gradually later in the day. The kittiwake's fishing trips moved with them, so that by the end of the month it was starting fishing at 8 in the morning and ending in the early afternoon, and again from about 8 in the evening until midnight or one o'clock in the morning. As the neap tides were less energetic, and the concentration of prey was thinner, the kittiwake was also fishing for longer.

These may be slight and simple results, but the implications are significant. The kittiwakes were not simply wandering out to sea, hoping for the best, but making careful and strategic decisions, thinking not only of the likely location of prey but of the evolving daily pattern of the sea, knowing at its nest 25 miles away what the conditions were likely to be on the fishing grounds, adjusting for abundance of prey, working its way through the rhythm of the tides.

By 2007, GPS loggers had been miniaturized to the point where they weighed less than a third of an ounce, which a

1-pound kittiwake could carry without any apparent effects. The first to be fitted with them that year were nesting on the radar tower of a derelict Cold War tracking station on Middleton Island in the Gulf of Alaska. The Americans had used it as part of the early-warning system for a Soviet attack through Siberia, but by 1963 the cost of maintaining it was too much and the station was abandoned. Artificial nests were added to the derelict tower for a new kind of tracking to take the place of the old, and by 2007 almost a thousand pairs of kittiwakes were nesting there, most of them using the wooden ledges that could be observed from inside the building through sliding panes of one-way mirror glass.

The Danish bird scientist Jana Kotzerka and the German seabird ecologist Stefan Garthe reached through the windows of the radar tower and put trackers on fourteen of the breeding birds. After a few days, they were all recaptured. Two birds had got rid of the GPS logger by pulling out the feathers it had been stuck to. Two of the loggers failed to work and one bird, inconveniently, had sat on its nest for two days after the logger was attached so that the battery had run out before it had flown any distance. The scientists had records of sixteen fishing trips from nine birds. Nine of those trips were just offshore, no more than 8 miles or so. The other seven were at least 22 miles each way, often overnight.

The kittiwakes were again strategizing their lives: short fishing trips to bring food back to regurgitate for the chicks in the nest, probably focused on locally available low-energy plankton sources; long fishing trips to get food for themselves, almost certainly from higher-energy concentrations of oily fish offshore. There was nothing random about the kittiwakes' search for food. The birds were carefully dividing their time and effort between two different sorts of fishing. It was the first intimation that the kittiwake was capable of flexible, intelligent decision-making, allocating individual life-tasks to different parts of its seascape, understanding that the chicks needed a constant and regular supply of food; and that their own energy budget needed more than the waters local to Middleton Island could supply.

When you next come across a kittiwake at sea, think of this: it is fixed on a particular task; it has a goal and a means of achieving that goal, not driven by mechanical instinct but by a mind that can prioritize and allocate different days' flying and hunting to different purposes. If you come close to the bird and catch the look in its eye, perhaps as it hangs over you quizzically in the sternsheets, ask yourself again: what is the intention behind that look? whose world are you sailing in?

The same summer, a different kind of tracker, a geo-logger first developed by the British Antarctic Survey scientists which cannot locate birds as precisely as the GPS but whose power supply can last an entire winter, was fitted to the legs of eighty kittiwakes on the Isle of May off the east coast of Scotland. Maria Bogdanova from the Centre for Ecology and Hydrology wanted to know if there was any connection between a kittiwake's success as a breeding bird and where it went in the winter. The Isle of May birds were having a bad time: only one chick fledged that summer for every four nests. Most of the adults that were dying were dying in the winter. Could the new technology throw any light on their struggles?

Just under 40 per cent of the loggers came back to the Isle of May the following year and the results gave rise to more questions than answers. Almost every kittiwake (94 per cent) that hadn't managed to breed the previous summer set off in late July for the West Atlantic, staying for long parts of the winter to the south of Greenland. Almost as high a proportion of the kittiwakes that had raised a chick (80 per cent) didn't do anything of the kind but stayed in the East Atlantic, so that the winter gatherings of successful birds and unsuccessful birds from the Isle of May were over 1,250 miles apart.

What could this mean? Was it that the better birds were genetically programmed not to make the long journey out and back to the American waters? Was it that successful kittiwakes chose closer wintering areas because the sea there was richer, calmer and easier? Were the non-breeders actively shut out of the good eastern wintering areas by the tougher, stronger, more successful kittiwakes? Dr Bogdanova had no answer to these questions, but she had demonstrated for the first time another game-changing aspect of seabird life: there was a link between individual summer breeding performance and winter distribution. Again, the implications are huge: the ability of a seabird to breed does not depend only on its summer performance, its ability to catch the prey its chicks need; only if it can carry over good effects from a successful winter will summer breeding produce a fledgling. Wherever they happen to breed, the entire North Atlantic is implicated in the well-being of seabird populations. There is no such thing as local here.

As this use of the new technology starts to mature, the results become increasingly nuanced and subtle. Between 2008 and 2011, Rachael A. Orben and her colleagues tracked ninety-nine black-legged kittiwakes breeding at St Paul and St George in the Pribilof Islands far out to the north of the Aleutians in the Bering Sea to the west of Alaska. The birds travelled vast distances, wandering over the deep oceanic waters of the North Pacific, some to California, others up into the Arctic Ocean, some to Kamchatka and Japan, one or two almost to Hawaii, covering an ocean province about 16 million square miles in extent.

Orben's findings clearly suggested, as others had, that birds could remember where they had been the year before and often went back there. This was no rigid pattern and the kittiwakes shifted from one season to the next, staying longer where the sea temperature was colder, homing in on ocean fronts where big water masses met in the sort of dynamic boundary at which sealife concentrates. Flexibility was part of what they were, and individuals were clearly able to explore and use many different kinds of winter ocean during their life-times. They were, in fact, much more adaptable and variable in where they went than the puffins that Tim Guilford's team had tracked from Skomer. Orben thought this was probably a reflection of the two different birds' life-skills. Kittiwakes are far better flyers than puffins, able at small cost to travel immense distances on their long and buoyant wings. The short-winged puffins have to expend high amounts of energy flying towards their destinations. But puffins are much better fishers than the kittiwakes, able to dive to 200 feet or more whereas kittiwakes can dive no more than a foot or two into the surface waters. Good flying wings allow the kittiwake to get to where the prey is at the surface of the ocean; great diving ability allows the puffin to get down to the fish or plankton wherever the puffin wings are able to take it. They have divided the dimensions between themselves, one a specialist in the horizontal, the other in the vertical, and those choices allow both to proliferate in their millions across the northern oceans.

*   *   *

All is not well. Across those northern oceans the kittiwakes are dying. From the 1970s onwards, which is when large-scale monitoring of seabird numbers began in earnest, the kitti-wakes have been showing perplexing evidence of being in trouble. Colonies all over their range from the Bering Sea to Arctic Canada, Greenland, Norway, Spitsbergen, Iceland, the Faeroes, Scotland, Ireland and England have suffered in some cases cataclysmic declines. In the United Kingdom as a whole, there was a 60 per cent decline in the thirty years from 1986 to 2016. Many places in Shetland which once had booming and shrieking kittiwake colonies are now silent and empty. Declines that had begun in the 1980s are continuing today. The island of Noss has lost 93 per cent of its kittiwakes since 2000, Foula 86 per cent, Fair Isle 90 per cent. In Orkney at Marwick Head 91 per cent of the kittiwakes have gone. In 1987, there were 7,829 kittiwakes on St Kilda, out in the Atlantic to the west of the Outer Hebrides. In 1995, there were 513 occupied kittiwake nests there, with fifty-six chicks born. In 2016, there were four occupied nests, a 99.2 per cent decrease, and a single chick that fledged. The colony at Hornøya sank from 21,000 pairs in the early 1990s to 9,000 twenty years later. The numbers in Spitsbergen have dropped as fast. In Alaska, the colony on Middleton Island is in dire trouble: it has crashed more than 90 per cent – from 83,000 pairs to 6,200 pairs – over the last twenty-six years. In the many colonies scattered around the islands and headlands of Prince William Sound, counts at colony after colony have been returning disastrous results for years.

The picture is more complex than those figures might suggest. Although something like 10,000 are still shot by the Greenlanders every year, globally, there are probably still enough kittiwakes for there to be no concern for their future. They are not on any red list for extinction. They certainly went through a boom in the course of the twentieth century, perhaps peaking in the 1960s, precisely because laws were passed in most developed countries against killing them for fun or hats. The recent decline may be a readjustment in the wake of that boom.

Where there is a problem, it is in the failure to breed. By 2008, only one nest in every four in the British Isles was producing a kittiwake chick that fledged. In 1986, on average, every nest had produced a fledgling. In the Pacific, breeding success has been lower than that since the 1970s. The parents have not been able to find the fish with which to feed their young, and that question of flying time may be at the heart of the problem. In Shetland, in a season when almost all kittiwakes were feeding within 3 miles of the colony, 81 per cent of the young fledged. In other years, when no young survived, feeding areas were up to 50 miles away.

Kittiwakes may look like the embodiment of flexibility, beautifully afloat on the vagaries of wind and ocean, but there is a rigidity here. Only if the fish – and they usually rely in midsummer on the newly emerged sandeels or capelin – are within an economic catching distance can they feed their chicks. These shoals are mobile but the nesting cliffs are fixed, and at times the ways of the ocean may mean that the link between nest and fish stretches too far. Sometimes, the season begins well, the birds feed happily, building nests and laying eggs, until something goes wrong. In Shetland in 2001, the first chicks hatched towards the second week of June. Only then, and quite suddenly, as Martin Heubeck, the Shetland bird ecologist, has recounted, 'chicks began to die in the nest, and most Shetland colonies had failed completely by early July. Food shortage was undoubtedly the cause, particularly the lack of sandeels.'

In the summer of 2011, a team of French and Norwegian bird scientists compiled what was, in effect, the chronicle of a catastrophic year on the island of Hornøya in the Barents Sea off the northern coast of Norway. Along with much of the kittiwake population of the Atlantic, the birds on Hornøya had been struggling for years, chicks failing to fledge, the population dropping. Aurore Ponchon from the University of Montpellier wanted to know why.

It is now well established that when fish are thin on the ground, seabirds can hunt for longer or go further afield or after different fish. Kittiwakes, when there are no fish to be had, can eat plankton or hang around fishing boats waiting for discards or even slumming it at fish factory outflows for the offal. But when there is an egg or a chick on the nest, those options may not be available. The adults must return to defend the nest or sit on the egg or feed the chick. Across many species of seabird, it has become clear that as long as a third of the usual amount of prey is available – and hasn't moved away or been fished out – the birds will be fine. That is a buffer built into their survival system. If less than a third of what they might expect is there to be had, they will be in trouble. That proportion is a critical threshold, after which the graph heads for the floor. Kittiwakes at Hornøya feed mostly on capelin. Only when the capelin are absent do they turn to herring or small shrimps, and the years in which they do that are the ones that are bad for breeding. The colony had yo-yoed and zigzagged over the preceding decades, occasionally more than half the pairs raising a chick, sometimes sinking to no more than one in ten.

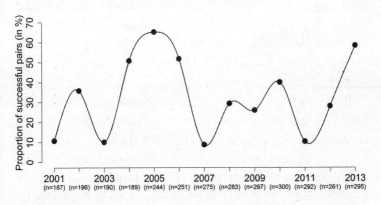

The ups and downs of Hornøya kittiwakes: good years and bad years
for raising chicks

To begin with in 2011, all was well. Most of the birds laid eggs, the capelin were just offshore throughout May and the kittiwakes were keeping themselves in good heart. By the first weeks of June, more than 80 per cent of the couples hatched a chick. For the next couple of weeks, most of the kittiwakes managed to keep the show on the road. About 10 per cent of the birds lost their chicks in that time, but nearly three-quarters of the pairs reached 19 June with at least one chick still in the nest. But the parents' behaviour had changed. It was obvious that something had gone desperately wrong with the supply of food in the sea. In May, the birds had never gone more than 12 miles from the colony on their fishing trips and usually for about two or three hours. Now though, with young chicks to feed, they did one of two things: they went off somewhere away from the nest but still on the island for half an hour or so. The scientists interpreted this behaviour as the parent getting away for a while from the anguished importuning of a hungry and demanding chick. Those parents also started

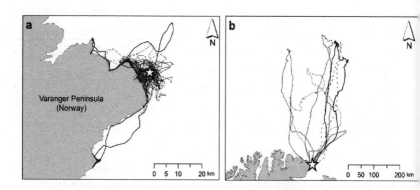

What happens when the fish leave: a) Kittiwake fishing trips from Hornøya when the capelin were just offshore in May 2011; b) the scale changes by a factor of 10: Kittiwake fishing trips from Hornøya in June when the capelin were driven north by warm Atlantic waters coming from the south.

to make very long trips away to the north, fifteen times the average length of their previous journeys, lasting up to forty-eight hours and with a total flying distance on each trip of up to 300 miles, deep into the Arctic Ocean. No scientist had ever recorded kittiwakes making fishing trips of such enormous length during the breeding season. As they were desperately looking for food for those chicks, their energy expenditure reached a level that was four times what it would have been had they been sitting on their nests.

The southern Barents Sea is an important nursery for capelin. In warm years, of which 2011 was one, Atlantic water floods in from the south and the capelin are wafted northwards away from the Norwegian coast. That oceanographic switch was the catastrophe dumped all over the lives of the kittiwakes. Until 19 June, on these long, expensive journeys north to the colder water and the fish, the adult birds were trying to keep the breeding project going. But on that day, the equations stopped working in their favour. Young chicks cannot absorb too much food at once and need to be fed small amounts almost every day. Such long, energy-consuming fishing trips could not deliver what was needed and were putting the parents themselves in danger. The graph had reached a threshold, and on a steepening curve most of the kittiwakes gave up the effort to feed their chicks. By mid-July, over 90 per cent of the chicks had died, deprived of food by the ocean currents. In some parts of the colony, only 3 per cent of pairs fledged a chick.

In a classic study in the late 1970s, Barbara Braun and George Hunt, from the University of California at Irvine, went to the Pribilof Islands in the Pacific far to the west of Alaska. Whole nations of seabirds breed on these volcanic islands – they have been called the Galápagos of the north – including over 30,000 black-legged kittiwakes. Braun and

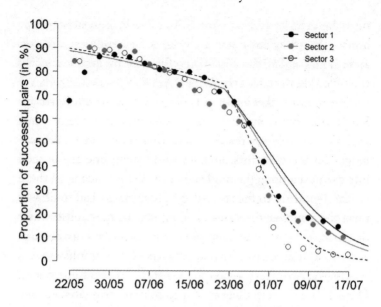

Threshold day, 19 June 2011, on which the graph turned; the
Kittiwake parents on Hornøya could no longer sustain their chicks'
needs and the chicks started to die. All but 8 per cent had died by
a month later.

Hunt knew that the kittiwakes on the Pribilofs usually laid two
eggs but almost always raised only one chick. No one had yet
established why.

They set up camp to examine the colony, weighing eggs,
ringing those chicks they could reach within twenty-four
hours of hatching, weighing chicks every other day and staring
at them from about 15 feet away, for up to thirteen hours at a
time. The story was Shakespearean: the first kittiwake chick
hatched from a bigger egg and a few days before its sibling, so
that by the time the second chick appeared, the first was
already 25 per cent heavier, dominant and a bully. As soon as
the younger chick appeared, its elder sibling started to soften
it up with what Braun and Hunt called 'intimidation, monop-

olization of parental feeding and beatings'. Within twenty-four hours of hatching the younger chick was being shaken, dragged and pushed around the nest, pecked and bitten. Within a week, the older chick had killed it.

The method the killer chicks used was simple enough. Kittiwakes build small nests of seaweed and grass stems gummed together with mud and droppings on to narrow ledges on the cliff. Bits of string and fishing line are pushed into the nest among the mud and grass. Below each nest, there is a straight drop to the sea. All the older chicks had to do was push the younger ones out of the nest and they fell to their deaths, usually weak but always alive as they fell. Between half and two-thirds of all second chicks were killed in this way.

Kittiwake fratricides went up when the weather got worse. Only if the fish were there and the first chick was satisfactorily fed could the second chick hope to survive. The sea speaks to the bird through the chemicals of its own body. As other studies have shown, if the lack of nutrients in the bird's body raises the level of the stress hormone corticosterone, that can act as a signal in the older that cruelty would now be useful, not only to itself but to its parents. Killing off the weaker chick in a year of dearth will make the stronger chick's likelihood of surviving much greater: it will get all the food. Only in a year with plenty of fish is it worth the parents' effort to feed both of them. Raised corticosterone in the blood – which both chick and parents experience – is not only a signal of distress; it is an instruction to abandon the second chick and hope for better luck next year.

The break point in Aurore Ponchon's graph on Hornøya 'suggests that kittiwakes crossed a threshold beyond which they stopped investing in reproduction, leading to extensive chick mortality'. The rising level of corticosterone in their blood had made them try harder, go further and expend far more effort. They had also maximized their resources by spending time away from the nest to 'avoid continuous chick begging', behaviour which is made more insistent by raised levels of corticosterone in the chicks' blood, to the point where the parents could not bring themselves to continue. Their stress levels will have instructed them not to go on and so the chicks died either because their older siblings killed them or because their parents decided to starve them in order to save themselves for another year.

This might look like a small Nordic tragedy – impoverished fishermen-farmers desperately trying to keep going in the face of a hostile world – but it might also be seen as a highly evolved and effective means of living with an inevitably variable resource. Ocean currents and the fish and plankton that live in them are unpredictable. They will only ever deliver probability not certainty, and so a breeding system which plans for the best but responds to the worst, reaching hormonally defined tipping points, at which abandonment or siblicide come to look like better options than maintenance or fraternity, might be seen as a system as exquisitely evolved as a lily or a rose. There would be no beauty in a kittiwake if kittiwakes did not exist, and their existence is predicated on exactly these strategies and choices of – occasional – starvation and murder.

It may be that the failure to breed in an individual colony and the steady death of adult kittiwakes is already accounted for in the workings of the larger system. It might be easy to accept that regular death and collapse are part of how things are in the world of the kittiwake, part of the system for survival and expansion this beautiful gull has devised for its life on the northern oceans. On the Shiants, the numbers have dropped, from 2,000 pairs or so thirty years ago when I first became conscious of them to about 500 pairs now. At those Shiant colonies, in which the armour-piercing, almost peacock-screaming cries still ring around the basalt cliffs, you would not think anything diminished if you did not know. They continue to look and sound like one version of the perfection of life. There have also been years when the kittiwakes have started to build their nests, but just as on Hornøya, or in Shetland, the fish have failed, the kittiwakes have abandoned their chicks and left to return another year. That absence and that silence is devastating at the time: the cliffs in those empty summers are not white with kittiwake guano but browning, returning to lifelessness, a suggestion of what life would be like if all seabirds failed, perhaps of what life might be like at the end of this century, when by some estimates most of the world's seabirds will have gone.

Aurore Ponchon performed one further fascinating experiment in 2011. She deprived some kittiwakes of their chicks and then fitted the adults with trackers to see what they would do in response to their (artificially induced) failure to breed.

She put some very expensive (£1,500 each) Platform Terminal Transmitters (PTTs) on three kittiwakes, one female and two males. The virtue of these transmitters is that they do not have to be recovered for their data to be collected. They feed their information direct to satellites. They are the instruments with which to follow birds that are leaving their old lives behind. Immediately after their eggs hatched, Ponchon removed the chicks, so that the adults would consider they had failed as parents. While incubating the eggs these kittiwakes had been fishing within 25 or 30 miles of the island. But deprived of their chicks, they changed their pattern of life, prospecting far and wide up to the north and to the Arctic islands of Spitsbergen, pursuing the capelin but also looking for a place in which breeding might be successful in the future. Her two PTT-generated maps – one localized, the other oceanic – are an epitome of everything which the life of the

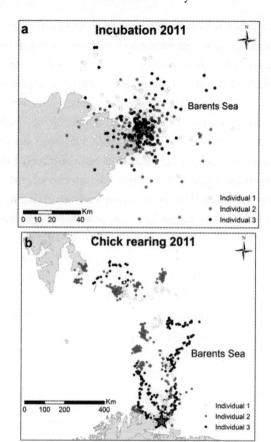

Hornøya kittiwakes a) going fishing nearby when incubating the eggs;
b) exploring the distant north as far as Spitsbergen 500 miles away,
once their breeding attempt has failed and they must look for improved
chances elsewhere.

kittiwake can add up to: attending to the eggs and chicks at the
colonies where we see them; roaming the world ocean, free
and beyond any human sight of them.

This is resilience on a map, the ability of an organism to
respond to challenge, to embody change in its own life, to find
a way through the labyrinths of the ocean. It is the one aspect

of kittiwake life which, in the face of the deep distortions to the world ocean brought about by global warming, is perhaps a source of hope.

But it raises a set of questions: is the world changing too fast for the resilience mechanisms of the birds to respond to it? Have the timescales of bird and world got out of synch? The kittiwakes, as surface feeders, with an ability to cover huge distances but not to penetrate beyond the surface of the sea, may be the most acute and sensitive barometers of just what is happening to the sea. The irony is that the very civilization that has created the instruments by which we can now track the birds – technological, industrial, energy-consuming, satellite-launching – is at the source of the changes which are making life on the ocean difficult and hostile for them. There is elasticity in nature, but is it elastic enough to cope with the strains we now at least know we are imposing on it? Does the kittiwake and its fate concentrate the dilemmas of the future?

4

Gull

The new island of Surtsey broke surface in the sea to the south of Iceland in November 1963, expelling steam and ash in a volcanic plume 2 miles high, glowing at night as a beacon of new creation. The great rift down the centre of the Atlantic had once again given birth. By the beginning of December, only two weeks after the eruption had begun, gulls were seen touching down on the steaming island, even as it was coughing and belching into existence. Two satellite volcano-skerries, Jólnir and Syrtlingur, both now washed back into the sea, appeared off Surtsey, erupting almost continuously and for two or three years coating it in a thick layer of volcanic ash. Every day, the North Atlantic winds picked up that ash and blew it in a gritty, intolerable sandpaper fog over Surtsey.

No birds could nest in those conditions, but thousands of great black-backed gulls and kittiwakes, most of them young prospectors, used to alight on the stiffened lava, up to 10,000 at a time, perching on this new roost in the ocean. Oystercatchers and turnstones started to walk on the black sand beaches. Red-necked phalaropes picked for food in between the floating rafts of pumice offshore, perhaps because the pumice had a calming effect on the sea. Surtsey continued to gulp and jerk into the ocean, but the birds were already clustering around it. Between eruptions and even during them, kittiwakes used to come to land on Jólnir and Syrtlingur. Wheatears danced around the edges of the new lava flows. Gulls and ravens sorted through the carcasses of dead birds and mammals washed ashore, feasting on the planktonic krill that in summer came to lie in pink masses on the sand.

It was a chapter from a marine Genesis: a new piece of earth, delivered from the depths, colonized by the creatures with wings. It took time for them to settle there. Not until 1970 did birds actually nest on Surtsey, one or two black guillemots down at sea level in tiny hidden cracks in the lava and a couple of pairs of fulmars perching on shelves and corbels on the rapidly eroding cliffs. These two species were alone until 1974 when a single pair of great black-backed gulls bred there. Slowly the ensemble gathered: kittiwakes and Arctic terns in 1975; and in 1981 herring gulls. By 1990 numbers had grown: 129 pairs of fulmars and 126 lesser black-backed gulls. In 2005 the first puffins made a burrow on Surtsey. A few decades after it had emerged from the waves, the birds had made the new land their own.

Above all, it was the gulls that had transformed it. For the first few years the new island's surfaces were mineral, planetary, almost indistinguishable from a dark, distant moon. Only on the upper parts of the beaches, where seeds of sea rocket and other salt-tolerant plants had been washed ashore, was there any vegetable life. But the gulls are omnivores; they feed on fish and eggs, scavenge what they can, take offal and discards from fishing vessels while preying on other birds and stealing from them. Of course they defecate on land, and regurgitate what they cannot digest. Some food gets half eaten; the rest is left to rot. Remains and pellets found at gulls' nests on Surtsey have included a leg of a gull and other bones, the head of a razorbill and legs of a young unidentifiable auk. Puffin and guillemot eggs from nearby islands, bits of gull chick, part of a snipe and odds and ends of pollack and other fish have all been found where the gulls were nesting. Fulmar eggs, which stand out big and white on the black volcanic rocks, were the most obvious of targets for the marauding gulls.

It was all fertilizer, nutrients pulled in to make a dead world live. After the gulls had set up a big colony in 1984, Surtsey greened up. Grasses, dandelions, hawkbits, ferns and moss all blew in to establish themselves in the gull-enriched ground. Snow buntings, which need seeds and insects, only arrived to nest in 1996 when there were enough plants for them to live on. In 1998, the first bush was found on the island – a tea-leaved willow which can grow up to 13 feet. Greylag geese, which are grazers, followed the establishment of big areas of grass. Meadow pipits, which usually put their nest in the side of grassy tummocks, followed them. Ravens, thieves and robbers, arrived to raise young on Surtsey in 2008, by when a good bird population was there to provide them with a living. A steady nutrient transfer from ocean to rock, largely facilitated by the gulls, had created a world in which other creatures could live.

Gulls are the early adopters, unlike the birds of the true ocean, which tend to be careful and conservative, keeping to what they know, breeding where they were born, long lived and late to mature. Gulls are different, on the edge of the seabird world, coastal creatures, living on the ecotone, that margin between life systems, picking at the leavings of the tide, relishing the comings and goings of a beach. They are not unlike us, who have always thrived on the shore, shuffling our way through its multiple resources, turning to the sea when the land is inadequate, to the land when the sea refuses to deliver. We and the gulls are co-habitants of the same world, uncomfortably recognizing each other, thriving in the same way, failing in the same way, behaving badly in the same way.

There are thirty-four species of gull in the world and, for all the differences in size and colouration, they are closely similar in life-habit and display, in their relations with each other, in what might be called their mentality, their opportunism, their particular mixture of the brilliant and the obtuse. Unlike most seabirds, the gull will lay three or even four eggs. If the eggs are removed, they happily lay more, aware that in the threatened, usually unprotected environments of their nests conservatism will not win out. They live on the edge, throwing resources at a problem, booming in favourable times, crashing when the world turns against them, gamblers and opportunists, bouncing and prodding their way through life.

They are not as clever as crows or parrots. They cannot, like pigeons, tell the difference between paintings by Picasso and Monet, Chagall and Van Gogh. And they can't count. When the pioneering American ornithologist Reuben M. Strong, then an instructor at the University of Chicago, went to examine the life-habits of the herring gulls nesting on an island in Lake Michigan, he found a tent indispensable.

> On approaching a breeding colony of gulls a wild panic
> begins which does not cease so long as the intruder appears
> to be in the immediate vicinity. I had a tent made … about
> four feet high and six feet long at the bottom. The cover was
> made from dark green cambric lining cloth, costing seven
> cents a yard. The entire cover could be folded into a package
> small enough to go into a coat pocket.

But simply having a tent was not enough. If he set it up on his study island and went into it alone, the gulls did not come near, all of them settling at the furthest point on the island away from the tent, or even afloat in the water near the shore.

Unlike the true birds of the ocean, whether the auks or the fulmars, herring gulls know that people spell danger.

> It was an hour and thirty-five minutes before any came close to the tent, and not until half past four the next morning, or thirteen hours after setting up my tent, did the first one alight on it. The gulls on this occasion were quite uneasy and were frequently thrown into a panic by their own actions. The sudden alighting of a bird, or a fight between two, would frighten one, who, not waiting to see what was the trouble, would take wing, followed by one after another as the panic-formed wave swept over the island, leaving it almost bare of gulls. Fortunately they quickly recovered and returned, but their alertness and the frequent 'wak-wak, wak, wak' of their note of suspicion showed that something bothered them.

Strong soon found a way round this. He took a friend with him into the tent. After a few minutes inside, the friend walked out again, went down to the boat, left the island and did not come back until the evening. A moment's pause and the gulls were soon happy: they were now alone on the island! With an empty tent and no one to bother them. Strong could spend the whole day watching them within a few feet of his hide.

Now and then, a gull seems to be cleverer than you might think. On a summer afternoon in July 2005 in the Tuileries Gardens in Paris, people as usual were feeding bread to the ducks in the big stone pools laid out in the seventeenth century. At one pool, where a couple of bird scientists happened to be watching, a herring gull arrived, rushed into the gang of mallards, grabbed a piece of bread, swam a few feet away towards the middle of the pool and then started to lift and drop the bread into the water. Each time, the bread broke into smaller pieces. Once he had scattered it into perfect crumbs, the gull stayed there waiting, as still as glass, neck out, head down, looking intently at the surface of the water. After a few seconds, the goldfish in the pool came up to nibble at the crumbs and the gull struck, about one in every two strikes resulting in a catch and a goldfish down the gullet. Once the bread had gone, the gull swam over to the people around the edge of the pool, stared hard at them, waiting for another piece of bread to be thrown in. After they had given it one, the gull repeated the whole process. It never ate the bread.

That is the only time a wild gull has been seen to cast its bread upon the waters, but all gulls are opportunistic omnivores, looking for food in all kinds of ways, scavenging, stealing and hunting, no species of gull specializing in a particular kind of food. Anything available will do. In the heroic Arctic world of Spitsbergen, the large glaucous gulls, called in Latin *Larus hyperboreus*, the super-northern gull, pale, ethereal and giant versions of a herring gull, hunt birds for a living. Their favourite is the fat and chunky little auk, the smallest of the Atlantic auks, like a compressed razorbill or a puffin canapé. It is a perfect bundle of fat, sitting like a sausage on the water, so that the coming together of little auks and glaucous gull looks inevitably like the meeting of greed and food.

The gulls feast on the little auks in any way they can, as eggs, chicks, young birds and fully grown adults. In the spring, as the Spitsbergen snow is melting around the little auk colonies, the glaucous gulls, in their beautiful pale coats, one of the ultimate birds of the north, sit on the few remaining patches of snow waiting with their heads down so that the yellowness of their bills and the brightness of their eyes cannot be seen, settled carefully into a camouflage ambush. As the little auks land, or even as they fly low over the colony, the gull jumps from the snow patch and grabs it and, like all gulls with all meals, swallows it whole with scarcely a thought beyond that look of mild alarm in its unblinking eye.

It is difficult to guess what is going on in the gull's mind from that story, but there must be elements of understanding there: the gull needs to imagine what the little auk is thinking and how it will behave; it needs to know that its own feathers will provide camouflage on the snow patch and that if it keeps its head lowered it will be less likely that the little auk will see him. Whether those forms of understanding are conscious seems unimportant. What do we know of what we do?

Canniness is everywhere in the glaucous gull world. Cruising over the Magdalenefjord in Spitsbergen, where the glacier tongues come down to the milk-green sea, they are on the lookout for the little auk fledglings. Some of the fledglings are separated from the parents that have been guarding them. But even these naive young little auks know how to dive as a form of protection and gulls cannot plunge beyond a foot or two into the water. As Dariusz Jakubas and Katarzyna Wojczulanis-Jakubas from the University of Gdańsk have explained, the gulls go carefully, slowly flapping high above the fjord, surveying the territory. When they spot a little auk fledgling they do not drop on to it immediately, as the fledgling would dive, but fly on for a short distance as if unconcerned, pull up and then glide back, no fuss, no wing gestures, but accelerating, dropping into their own speed dive, so that when only a few inches above the surface of the fjord the gull can approach the little auk hard, fast and silently, with no warning, and snatch it on the wing. This stealth-bomber approach-flight, which involves thought, an understanding of little auk behaviour and of concealment, the role of speed, body-stillness and deceit, ends up with a dead little auk four times out of ten.

Finally, when the surface of the fjord is covered in small icebergs, known as 'growlers', and other pieces of ice, the glaucous gulls swim quickly towards the fledglings in a series of zigzags, lowering their heads as they come close and at the last minute snatching the prey. The Jakubases think the zigzagging is designed to confuse the little auks, which cannot guess where the attack will come from, and may not be able to distinguish between 'a white-plumaged gull not swimming directly toward it and the slow-moving, snow-covered ice fragments'. This is Homeric ambush warfare, gull as a master of deceit, one of us.

We may need a new definition of intelligence in animals. It does not consist merely of tool-making, adapting bread as fish bait or curling leaves to extract ants from a hole. It makes sense as a more comprehensive category, running from predatory canniness to body-design, from aerial skills to the form of the feather. Whatever term you want to use – intelligence, capability, admirability – this quality of excellence in the bird extends all the way from an inventive form of hunting to ingeniously adapted mating behaviour, to a resilience and responsiveness in a changing world, to an effective endocrine system. The whole seabird is the vehicle of intelligence, its memory, its eye, its colouring, its aggression, its fear, its beauty.

One of the great twentieth-century bird discoveries was that behaviour may last longer through evolutionary time than the shape of the limbs; shared habits may be a better pointer towards a shared ancestry than similar bodies. The nineteenth century had expended extraordinary efforts in classifying and taxonomizing different dead bodies, birds that had been shot for the great museum collections of the industrialized world. The twentieth-century recognition, of which we are still the heirs, was to see that the living bird is more true to the bird than the thing in the morgue.

At the urging of the great animal behaviourist Konrad Lorenz, his heir and colleague Niko Tinbergen, with many others at Oxford, embarked in 1949 on a groundbreaking study of the gulls. Tinbergen wanted to understand 'why each species has the characteristics it has'. Such a simple question; such a revelatory set of answers.

The group asked, for example, why so many gulls in so many populations across the world have white heads, white chests, white bellies and white tails. How come that colour hasn't altered in all the varying circumstances in which gulls now live? It turns out that the whiteness is intimately connected to hunting behaviour. If you sit in a small boat gutting fish, throwing the guts into the air or dropping them in the sea beside you, the gulls will soon be with you, plunging down from a height, diving, braking and turning in the frantic, animated way we know, and exposing only their white parts to you.

The whiteness is in fact 'aggressive camouflage' – it allows plunge-diving birds to approach fish more closely than if they were dark. This effect was shown by some of the experiments made by the Tinbergen team: fish swam away more quickly and more energetically with a model of a dark bird hovering over them than with a white one. In a pool full of sticklebacks, the fish under the dark bird moved a full foot before the fish under the white bird had moved at all. This whiteness turns out to be so important that even in gulls that have dark upper wings – the black of the black-backs, the dark grey of the herring gull – the leading edge of those wings, the first quarter-inch which will also be glimpsed by the prey in the micro-second they have to survive before the gull gets them, are always white in all species. The only exception is the dark of the wingtips, where there is a trade-off between the strengthening effects of melanin and the alarm the blackness might give to prey. But even there, intriguingly, on the tip of the very first primary feather, the black is broken with a panel or 'mirror' of white.

There are some revealing exceptions. Why do young herring gulls not turn white until they are three or four years old? They remain a speckled, owl-like brown-white. It became clear, as soon as Tinbergen's pupil the Oxford ornithologist G. R. Phillips investigated them, that young herring gulls go out to sea much less often than the fully grown birds. Coastal flocks on both sides of the North Sea contained on average 30 per cent young dark birds, but only a mile out to sea that dropped to about 3 per cent. The young ones get their food by scavenging along the shore at low tide and so have no need of white chests and bellies. The speckled plumage probably plays some part as camouflage, protecting them from sea eagles and skuas.

Birds which don't plunge down from the white sky above, but chase their prey deep in the dark of the sea – the shags or cormorants or guillemots – don't need white faces, and would in fact be disadvantaged by them, and so they are dark headed. An exception is the guillemot in its winter plumage, when it has a white neck and cheeks. Is that a reflection of a change of winter lifestyle, when away from the colony it fishes for different prey?

Intriguingly, some gulls also have dark faces. Tinbergen and his team investigated these anomalies and it turned out that the black-headed gulls, which acquire that black head only in the breeding season and lose it in the winter, are largely land-based then and eat insects and larvae: no need for a white face because the grubs and insects are not on the run. Only in the winter, when black-headed gulls do go fishing, does the black mask disappear. But why bother to have a black face at all? Because, as the Tinbergen team found out with experimental dummies, black faces frighten other gulls. Turning a black face towards another gull keeps it at a distance, and so the black mask of the black-headed gull is a way of spacing out gulls in a breeding territory, a method which probably preserves eggs and chicks from the depredations of their neighbours.

But what happens when two black-headed gulls, each equipped with the terrifying face, need to meet and mate, to create the next generation? Straightforwardly enough, black-headed gulls, when establishing a bond with their mate, *face away* from each other, showing the white body, concealing the black face, getting close enough for other bonding systems, based on smell and touch, to take over. In this way the evolution of the gull has allowed two opposing behaviour systems to interact: social attraction, and the comfort and protection derived from living in a colony, brings them towards each other; territoriality, which protects them from each other, means that when together they are not that close. As some extraordinary observations of herring gulls and lesser blackbacks have shown, that protection from the neighbours can be of terrifying importance.

The world of the seabird is a universe of mystery, beauty and wonder; it is not, beyond the necessary parental duties, one of care. The cruelty and violence can sometimes be so pervasive that gull existence looks like a version of hell, one in which goodness does not have a part to play.

In the mid-1960s, Jasper Parsons, a young research student then at Durham University, set out to study the population dynamics of the herring gulls on the Isle of May, the rocky island off the east coast of Scotland. In three years between 1966 and 1968, he laboured to put more than 15,000 rings on the legs of his chosen gulls, first marking the young birds when they emerged from the egg, ringing them when they were a few days older.

But there was a problem: the chicks were disappearing. No sooner did he ring a chick than he lost it. Parsons soon realized why: the rings were surfacing in the regurgitated gunk of adult

Herring gull with kittiwake chick

gulls. He was surveying a population of cannibals. 'The Herring Gull *Larus argentatus*', Parsons began his paper in *British Birds*, 'is a notorious killer during the breeding season. Young chicks are repeatedly struck on the head or often gripped by the neck and worried until dead.'

Some chicks had only themselves to blame. They wandered into other gulls' territories and were killed there, usually by an adult gull giving them a fatal head wound. But in parts of the island things weren't so straightforward. On a promontory called North Ness, Parsons ringed 1,415 chicks in the early summer of 1968. Cannibal herring gulls living there ate very nearly one in four of them, 329 chicks, with each of the eaten chicks surviving, on average, less than a week. Amazingly, over half of these chicks (167) were eaten by just four cannibals. The habit was widespread, but it peaked in the impact of these four sociopaths. Each of the super-killers was eating 5 per cent of all the chicks hatched on the headland. Two of them lived on offshore rocks separated from the headland by cliffs and sea channels; from those isolated lairs they cruised out over the less dense patches of the colony looking for lunch, often taking the youngest and weakest of a clutch of chicks. They weren't alone in preying on their neighbours' children, but these four dominated the fate of the colony, and they repeated their predations year after year as Parsons ringed and watched.

They were not crazy adolescent birds. 'The cannibal gulls on the Isle of May', Parsons wrote sombrely, 'retained the habit and ate chicks in successive years: they were not unmated or immatures and, indeed, were no less successful at breeding than the rest of the colony.'

It had been suggested that gulls ate baby gulls to reduce competition for scarce resources. Parsons's conclusion was both calmer than that and more disturbing. 'Cannibalism need not indicate a food shortage,' he wrote, 'nor need it be an adaptation to curb recruitment. It is probable that, in a dense breeding colony, chicks are a more accessible and an economical or preferred source of food.' Chicks were tasty morsels, snacks for criminals, easily to hand. There were only four dedicated chick-eaters out of 1,800 birds – and yet chick-eating was clearly a reasonable response to the challenges of gull life. Precisely because it was a settled society, earning its income, creating the next generation and establishing homes, a marginal, parasitic, powerful sub-set of gulls could make their living off their own kind. Nice families were creating the conditions in which criminals could thrive. By definition, no colony could persist if all its members ate the next generation. And yet, in a colony of a certain size, a handful of cannibal kings could seem like an acceptable or even inevitable outcome, simply because there was no law there beyond the law of force. If a bird is a scavenger, why shouldn't it scavenge the neighbours' children?

Parsons identified and followed the career of one Isle of May gull that seemed to have stepped beyond even these realms of normal, rational, neighbourhood crime. This gull was a member of the North Ness four, but it is difficult to think of the way he was living as anything but a form of tyrannical madness, a gull Lear, a gull Pol Pot, acting out his lunacy and his terrorism on the subject creatures around him.

In the spring of 1968, this one gull ate over forty chicks while, alternating with his mate, also sitting on a clutch of three eggs of his own. When they hatched, the cannibal went on collecting chicks from the rest of the colony and bringing them back to his own nest, the downy bodies clamped carefully in his yellow bill. But now that he was beginning to raise his own chicks, he didn't eat his kidnap victims. Instead, for a week, he added them to his brood, which soon consisted of three of his own and eight he had stolen. In a rainstorm, two of them died of cold, one his, one borrowed. And that event seems to have released something deeper in him: he ate eight of the remaining nine chicks, including the two survivors of his own. One – the first stranger he had brought back alive – fledged, having been fed almost exclusively on baby herring gulls identical to himself.

What to make of this story, vertigo-inducing and gut-hollowing as it is? Had Parsons found an animal that was mad? It is a word scarcely ever used of wild creatures, perhaps because we want to imagine that nature does not contain madness. Or was the gull simply acting out the most fundamental of life-options, unconstrained by sentiment or custom, driven by selfhood and the urge to power? He was repeating the oldest myth of all, the destiny of Kronos, the father of the gods, who, sickle in hand, castrated his own father Uranus, as his mother, the Earth, had urged him, throwing the testicles in the sea (they frothed there and from the foam emerged

Aphrodite, the goddess of love). This root-god had many children of his own – Hades, Poseidon and Demeter among them – but Kronos (not Khronos, the god of time, but etymologically older, 'the god of the cut') was anxious that they might do to him what he had done to his father and so he ate them. Only later, tricked by another son, Zeus, did he vomit them back into the world, where they imprisoned him for eternity. Look at a herring gull now, sicking up food for its chicks, and you will have a glimpse of the foundations of the universe.

This is the first lesson of the seabirds. Beauty in the natural world is not sweetness and delicacy, although those qualities may often be part of it. Nor is it, necessarily, the play of power, the frisson of violence and domination enacted in front of you. It is this: the revelation of the actual, the undressed and naked presence of hidden principles – in this case the profound rivalry of parent and child – which the Greeks could allow to surface only in myth and which in our culture is scarcely allowed a presence at all. In some ways, the seabirds' cruelty is their beauty.

An uncomfortable question raises its head: are gulls naturally bad? Or does that question not make sense in relation to a bird? Is it stupid to apply moral categories to animals? Raymond Pierotti, an expert on the behaviour of both gulls and dogs from the University of Kansas, has made a study of the role of spite and altruism in gulls. They are a natural pair: altruism is kind, spite unkind, but neither by definition is done for a purpose beyond itself. Neither brings any advantage to the individual and both in fact may incur some costs. They are free-floating indicators of what an animal is actually like, beyond what could be called its usual economic activities of getting, spending, breeding and chick-rearing.

The question is complicated because a gull could do well from reciprocal altruism (I'll scratch your back if you scratch mine), from the way in which apparently motiveless spite can get rid of the competition – killing a rival for no good reason does also empty out the playing field – and because behaviour that might look altruistic may in fact be selfish in an evolutionary sense, at least if it benefits the individual's relatives more than it costs him. Evolutionarily, it makes good sense for a father to drown if in doing so he saves at least two of his own children.

Pierotti spent two years looking very carefully at the western gulls in their exceptionally crowded colony on the Farallon Islands off the coast of California – the most famous gulls of all: it was Californian western gulls that starred in Alfred Hitchcock's *The Birds*. Pierotti's first tests were in kindness. Could a gull be good to a chick that wasn't its own? Scientists are as unsentimental as gulls about their experiments and Pierotti 'removed' five female gulls and two male gulls from their nests on south-east Farallon while they were already bringing up a brood of chicks. The females deprived of their husbands tried to keep going as single parents but gave up after three or four days and left their chicks to die. No males offered themselves as second husbands. The bereaved males, though, all found second wives within two days, and without exception the new stepmothers looked after their predecessors' children, feeding and guarding them and in two cases sitting on the eggs that had been laid by the now-dead first wife to the point of hatching them. Were the women just nicer than the men? Maybe not: gulls are not particularly long lived – fifteen or twenty years compared with a gannet at over fifty and an albatross at up to eighty – but they do mate for many years, there is an excess of females on the Farallons and it may be that behaving as a good stepmother was an entrée into a marriage that the gull might otherwise have found difficult to achieve.

The king of gull love, though, is Franklin's gull. They breed in marshes in Canada and the United States and their eggs are laid on floating nests carefully moored to the reeds. In the summer of 1899, Thomas Roberts, a Minnesota doctor, went looking for Franklin's gulls on a marshy lake in Jackson County, at the southern end of the state. He found the nests of reeds and rush stems, 2 or 3 feet wide at the base, solidly made, with a more delicate superstructure, in the centre of which was a small dry reed-lined cup for the eggs, 10 inches across, 3 inches deep, well above the water. There had been heavy rains at the lake that summer, the water had risen and the nests 'were floating hither and thither at the mercy of the winds, but, strange to say, this state of things did not appear to greatly disconcert the owners'.

When the chicks are born, emerging as 'pink-footed, pale-billed little balls of down', some of them stay 'quietly in the home nest basking in the warm sunshine'. More often, though, and astonishingly:

> they are no sooner dry from the egg than they start to wander. A few are content to go no further than the broad sloping sides of the nest, and there they may be seen quietly dozing or tumbling about among the stems of the rushes as they explore the intricacies of their little island. The greater number, however, put boldly out to sea and drift away with the chance breeze, their tiny paddles of little avail as they pursue their now enforced journey. A gust of wind a trifle harder than usual, or a bump against a floating reed stem, and over they go, bottom-side up, only to come quickly right again, dry and fluffy as ever. Having after many failures crawled over the tiny obstruction, they sail contentedly on.

The adult gulls don't always welcome this behaviour and 'seizing the drifting chicks by the nape of the neck with the powerful beak, they are jerked bodily and roughly out of the water, and from a height of three or four feet thrown as far as possible in the desired direction'. But the extraordinary outcome of this marginal life is what Dr Roberts in 1899 called 'a curious spirit of communism among these Gulls'.

> An old Gull cared for whatever young Gulls fell in its way, and when the stray chicks chanced to clamber up into a strange nest, against which they happened to drift, they were after a few admonishing squawks welcomed as one of the household, and scolded, pecked, and fed just as though the foster parent had laid the eggs from which they were hatched.
>
> Now and then an entire brood would escape in a body and crawling up beside some incubating bird on a neighboring wind-ward [sic] nest would cuddle close about the old bird, who, to all appearances, was perfectly willing to adopt them in advance of the appearance of her own infants. Occasionally we saw old Gulls already in possession of a family twice the size to which they were entitled, rushing out and pouncing upon other fresh arrivals, who were quickly hustled and jerked up among the others until not infrequently ten or a dozen of these tiny balls filled the nest to overflowing, and in the diversity of coloration presented plainly indicated their varied parentage.

It is a world, Roberts says, of 'foundling asylums' in which chicks were fed without distinction.

Time and again we saw some little chap, just fished out of
the water and still sore from the rough usage to which he
had been subjected, fed to repletion by his captor, who
disgorged into the tiny maw a juicy mass of dragon-fly
nymphs brought from the meadow a mile away.

For entirely practical evolutionary reasons, this little noisy
black-headed, insect-eating gull has developed a community
chick service as its means of survival. There is, apparently, no
thought given to the welfare of the individual, nor do the birds
concentrate on those chicks that share their genes. So is this a
love culture? Or is it just another version of balancing the risks
and rewards of different survival strategies? If the entire colony
were seen as a single genetic cluster, this behaviour would be
merely self-interest masquerading as care, the genes in all
birds looking after all other birds. If all Franklin's gull parents
take in all passing chicks, the shared genes they embody will
have a better chance of survival, despite the costs to the adults.
For these inland gulls, unlike, say, the kittiwakes, multiple
chick survival may simply have turned out to be a better bet
than a focus on adult well-being.

The Canadian geneticist Paul Hébert has also found that
when gull chicks do not receive enough food from their
parents, they are programmed to run away from the parental
nest and look for adoption in a foster family, where
better-quality adults might give them the food they need. No
such thing as family loyalty in the gulls; all that matters is
getting on and getting out, using anyone and anything that will
help.

But these questions of love, loyalty, survival, dependence
and care all begin to look a little different when seen from the
point of view of the gull itself. If a gull on its nest sees a chick
floating past on the flood, or even finds a neglected chick

knocking on its door asking for food, then saves it, warms it, feeds it and nurtures it, how can we say that gull and chick are not bound together by love?

What about spite? After his years of observation in California and Newfoundland, Raymond Pierotti has one unequivocal conclusion: 'The spiteful behaviour that I observed was ... directed almost exclusively at chicks and was always performed by male gulls.' Seven or eight years after Jasper Parsons had witnessed the horrors on the Isle of May, two Cambridge researchers, J. W. F. Davis and E. K. Dunn, discovered what could happen if the habit of one or two individuals overtook an entire population. They began to look at the behaviour of another gull, the lesser black-backed, slightly smaller, not silver-plated like its cousin but darker on back and wings. Davis and Dunn's chosen domain was the wonderful island of Skokholm, a tiny seabird haven off the Pembrokeshire coast where the gulls were booming. Skokholm had always been a rich and fertile place, its farmer in the early twentieth century sending weekly baskets of eggs and tubs of butter over to the mainland. When he swam his cattle over there to market, pools of seawater used to lie in the rolls of fat on their backs as they walked up the beach.

The pioneering ornithologist and showman Ronald Lockley – for whom evenings on Skokholm meant a velvet smoking jacket and a double Dubonnet and gin, and who fed his pigs so many gull eggs that they came out in a rash – had made the island famous before the war as a dream escape from the rest of life. And he had never held back on the gull eggs, taking 3,000 a year from 1,000 nests to feed his family, pigs and friends. Eating gulls' eggs was banned in 1954 and after that the number of gulls rose fast. Even so, by the 1970s, it was becoming clear that lesser black-backed eggs

and chicks were for some reason failing to survive as each
season went on.

Sitting in their hide, peering out at the gulls around them,
Davis and Dunn soon saw what was happening: lesser black-
backed gulls were preying on their nearest neighbours, not
one or two crazed aberrationists but, as the summer
progressed, almost the entire population, systematically and
comprehensively. Nearly always the predator gull had lost its
own eggs just before turning on its neighbours'. A chain reaction built up of what the Cambridge biologists called 'aggressive failed breeders'. Loss of eggs made gulls take other gulls'
eggs, and so on in a domino run of social damage that spread
like a virus through the colony. Where bracken hid the nests,
the next generation had more of a chance, but the growing
density of nests – the birds were feeding on discarded fish
from the increasing number of trawlers in the Irish Sea – was
creating a new and intolerable pressure across the whole of
Skokholm gull society.

The irony, as Davis and Dunn pointed out, was that the very
tactics which the gulls had developed to protect themselves
from outside predators – nesting closely together and all laying
at the same time, so that any predator, however much damage
it did, would be swamped by numbers and be unable to kill all
the gull chicks – had now become their greatest vulnerability.
The enemy was within and the gulls had developed the social
version of an auto-immune disease, the whole population
gripped in a cascade of fatal aggression. The more closely they
nested and the more eggs and chicks that were available at one
time, the more disastrous the damage they could wreak on
themselves. The Skokholm gulls remain today in a pit of
self-destruction, numbers far down, in most years no more
than one chick raised for every thirteen nests, in each of which
the gulls lay three eggs or more. One fledged chick from forty

eggs laid: an entirely natural habit of self-destruction, the
apotheosis of spite, almost entirely performed by males.

* * *

Eighty or ninety feet above the seafront at Hastings, the
herring gulls are cruising on a slow winter morning, perhaps
two or three hundred of them on a half-mile stretch. About
one in four are adults, white and silver grey, *Larus argentatus*,
the silver-plated sea-mew. The others are at various brown-
flecked stages of immaturity, this year's fully grown chicks,
and many one- and two-year-olds, their eyes and beaks still
dark

A warm wind is sliding in from the Azores across the English
Channel. There's hardly a bird out above the shingle beach
itself. Most are just inland of that, above the tarmac and
concrete of the human strip in front of the Jasmine Thai
Cuisine café, the Regal, the Seagull, the Cod Father and Foyles
Pie and Mash Shop. The birds coast without effort in the wind,
up and down, kerb-crawling in the stream of southern air, as if
hungry for something or anything, attentive, prepared for
what human beings might offer.

They are the co-habitants of this town: cool, morose, not
quite at ease, bored and edgy, hung there as if part of a mobile.
They look well sprung, slowed down, a cartoon of small-town
criminality. Respectable citizens on the pavement; louche
parasites loitering above them. If birds could chew gum, these
would be.

I watch and listen to their bell-like crying and think how
beautiful it would be if we didn't know it too well, a high
summoning and warning, like bird muezzins. This is the
herring gull's long call, its neck down first, *hieeee*, then a
sudden jerking up, the head high, ululating, yellow mouth

wide open, its throat visibly oscillating with each syllable, the whole body gradually bending forward, a marshalling cry, *ay ay ay ay ay*, slowly lowering so that the gull ends nearly horizontal, its urge to call exhausted and all conviction gone. Then, as if the crisis were over, whatever it might have been, the settling of wings like the shuffling of cards and the bird preening its feathers.

The town is poor. The Victorian stucco of the front is greying and the grey-yellow joinery of the crescents is unpainted. As far as the gulls are concerned, the cliffs behind the buildings are continuous with the human world. They land on them and on the shit-strewn streetlights, and sit there scratching at their heads with one foot or the other. Some of their droppings fall to the pavement. Sometimes one of them comes down to look at me, scarcely a wingbeat as it slides over the clock golf and the turning sails of the mini-windmill. Beams and beads of light stand out in the Channel, but here all is gull grey. A slow investigative look, the head just cranked to one side, sometimes a single leg dropped below the body: chips, fish, anything? They are wary of each other and, if they cross paths, flutter and muddle in mid-air, as if a nerve shudder were passing through them.

At ground level the people hurry from one shelter to the next, shop to car to café, their bags and possessions tucked under their arms, hats down, not at home on these streets, nor alert to their environment, or what might happen next. A little spitting rain is coming in from the sea. Kenneth Hay and his wife Andy come quickly past, coated and hatted up, each with their fish and chips inside a polystyrene box, making for the car. They don't want to talk much. 'As soon as you sit down they harass you,' Kenneth says, looking at me as if I might be a gull. 'It's not pleasant.'

'They are vicious,' Andy says.

'Andy here had her fish and chips in one of these. She was at one end and the seagull was the other trying to get at it.'

Who won?

'It did. Or some of it anyway.'

'It's the noise I can't stand,' another hatted man on the front, Robert Cotton, tells me. 'Not first thing in the morning. And I have seen an old lady here walking along with a gull trying to pull her bobble hat off her head.'

On the municipal boating pond in Flamingo Park, rimmed in yellow bricks, about 150 gulls are standing hunched in water that is cobalt blue with disinfectant. The sun is lighting up their bright-blue feet. Three raggedy little red-legged turnstones, their backs the colours of a rough French terrine, are poking about on the brick and shingle island in the middle of the pond. They have come in here on their winter migration from the wildest places on earth, the west coast of Greenland or the Canadian Arctic. The gulls pay them no attention as the turnstones live up to their name, scrabbling in the shingle, nor does anyone else. I notice the little hedges of spiked barbs on the cornices and chimney pots of the buildings on the front, designed to prevent gulls nesting there. The gulls are standing on them.

In the course of the nineteenth century, seabirds were persecuted all over Europe and North America and their numbers crashed. Only after the Victorian legislation to protect them did they begin to recover. For the first three-quarters of the twentieth century, the herring gull population in Britain boomed, growing at 12–13 per cent a year, so that by their peak in 1970 a few hundred breeding pairs in 1900 had become 286,000, a total, if you add in the young birds, not far short of a million individuals.

Much of that was the natural bouncing back from the early Victorian killing, but there is no doubt it was pumped up by an increase in the rubbish we produced, the offal and discards from industrial fisheries and the landfill from our cities.

Herring gulls were first seen nesting on British buildings in the late 1920s, joined by their cousins the lesser black-backed gulls twenty years later. Until the 1940s most food waste was composted or burned at home, or fed to the pigs. Rubbish tips were filled with inert, ashy waste. The 1956 Clean Air Act, which prevented rubbish being burned on tips, started to provide gulls with an unlimited amount of food. A few of the largest gulls in the world, the great black-backed, with a wing-span approaching 6 feet, started to nest on British roofs in the 1970s. Street lighting allowed them to feed at night. But the introduction of black plastic bags, a perfect incubating environment for botulism, may have brought this growth in numbers to an end. Some wild colonies of gulls in the Bristol Channel dropped by 90 per cent in the 1970s to 1990s through botulism. But urban gulls seemed to have been less affected. When in 1990 an incinerator was installed at a rubbish tip near Brest in Brittany, and the amount of rubbish available to the gulls dropped from 5 tons to less than 1 ton a day, two things

Shiant gulls looking for scraps

happened. The average weight of the gulls in a colony on an island 8 miles to the south dropped by 5 per cent. But more significantly, the number of young hatched by each pair of gulls fell by a half. The implication is clear: the enormous volumes of twentieth-century rubbish had been creating baby gulls.

By the end of the twentieth century, the gulls had gained their foothold. Cities near the Great Lakes in America and on the Pacific coast, as well as all over Atlantic Europe, now had their gulls. One estimate in 2005 reckoned the number of urban gulls in Britain was growing by almost 25 per cent a year. This was against the background of a general decline in the numbers of gulls living in the wild, so that the million-odd herring gulls in Britain in 1970 had sunk to less than half that number by 2002. Two-thirds of them were now in cities and the herring gull had become a commensal with us, a table-sharer, as much part of our lives as house-mites or rats.

All over the world scientists have looked at how these urban gulls are doing and their findings do not make happy reading. The Australian silver gulls, smaller than the herring gull but related, usually feed on worms and fish, frogs and crabs, but in and around Hobart, Tasmania, as in most cities, they eat human junk. Heidi Auman and her colleagues at the university in Hobart compared these urban gulls, which had such an easy source of food to hand, with the same species living out on islands in the wilds of the Southern Ocean. The Hobart gulls turned out, as expected, to be bigger and heavier. But another set of studies, on northern hemisphere gulls, found that these urban gulls were big and fat at a cost.

For twelve successive years, Raymond Pierotti with Cynthia Annett from the University of Kansas looked at the western gulls living on Alcatraz Island in San Francisco Bay. Their gulls were in a kind of fertility sink: fewer eggs hatched, the food not good enough for usual chick development and fewer chicks fledging. The adults didn't live as long as their wild-food counterparts. Another of Annett and Pierotti's projects on a group of herring gulls found that the birds feeding on human rubbish laid fewer eggs, had fewer live chicks per clutch and fewer survivors from those that did hatch. A junk diet was making junk birds.

It's a disturbing realization: the gulls on the Hastings seafront, largely fed on fish and chips and on the rubbish from people's bins, have been transformed by the food we have been giving them. They are no longer wild creatures, but a reflection of who we are and how we live. They are the gulls of the Anthropocene. The impression you get, the instinctive gut feel of these birds on the seafront, that they are somehow degraded and impoverished, is borne out by the science. We have created a race of gulls that reflects the worst of us.

*   *   *

The industrialized world has been waging all-out war on the gulls for at least a century, largely because they cost too much. Gull strikes on aeroplanes cause about $40 million in damage every year to US civil aircraft. The damage to buildings costs far more than that, but the total has never been calculated.

A battery of weapons has been ranged against them, but with little consistent success partly because gulls are canny and learn to live with almost any threat, partly because of our own ambivalent attitude to them. We hate them but we don't want to behave horribly towards them. If we find them too insistently present in our cities or airports or even seabird colonies, we destroy them. But most control measures are made useless by this double handicap. Fireworks, rope-bangers, hawks, false hawks, model owls, balloons with hawk eyes, propane exploders, destruction of nests, the playing of the recorded distress calls (disliked by city-dwellers), mylar flags, overhead wires to catch gulls as they come in to land: all either work for a while only or have unwelcome side-effects.

Shooting them is difficult in cities and disliked by citizens. In Venice, the authorities give the pigeons contraceptives, but that doesn't work with gulls because they 'rush feed'. Some would overdose and die; the others would remain fertile. Methyl anthranilate, an irritant that makes gulls' legs sore, has been put in small, temporary pools of water. More horribly, the US Environmental Protection Agency has registered a chemical frightening agent called Avitrol. When gulls eat it, they become disoriented and emit distress calls while flying around erratically, frightening away the birds that are yet to be poisoned. Avitrol has been used at airports and those gulls that have been poisoned usually die within four hours. Poisons, such as alpha-chloralose, result in drowsiness and death over a few hours. Poisoned gulls tend to fly away and die inconveniently on pavements in towns. Citizens don't like that either.

Authorities have also gone for gull eggs, killing the embryo by vigorous shaking, pricking a small hole in the eggshell and injecting an embryonicide such as formalin. Eggs are porous, allowing oxygen to reach the growing chick. Spraying them with oil suffocates the foetus. In Maine, 800,000 herring gull eggs were sprayed with oil between 1940 and 1952, when the programme ended due to lack of money and the gulls resumed their natural course. Foxes and raccoons have been released into colonies in Massachusetts. In Alaska, pigs were used to clear gulls off an island next to a stretch of water used by seaplanes.

Does it have to be like this? Our relationship with the gull hasn't always been so hostile. Some seabirds that are good to eat, including the Manx shearwater and the oozingly fatty young gannets from Bass Rock, have been protected by law since the Middle Ages, but the first legislation in the world to protect any kind of seabird from too harsh a persecution, and not for food, was in the Isle of Man in the 1830s. Gulls were not to be killed because they acted as guides to fishermen looking for the herring shoals. Anyone who shot a gull at Port Erin on the Isle of Man 'either willfully or wantonly' would be fined £3, a month's wages for a working man in the 1830s. For Manx fishermen the gulls would show the way home when the compass was useless, and worked as cleaners in the fish-market in Peel, picking up the guts and offal from the market floor after it had closed for the day. Every morning the men would wait until they could see the flocks of gulls out to sea. Then they 'receive the signal with joy. Ever on the alert, they throw their nets in the boat, and when after the toilsome day they return laden to their homes, the auxiliary gulls receive the reward of their services in the small fry and garbage.'

I know men in the Hebrides who as boys kept seagulls as pets. 'We would get them when they were very very small,' Donald Macleod told me. He is a fisherman in Marvig, on the eastern shore of the Isle of Lewis, and because every second man in Lewis and Harris is called Donald Macleod, he is known to the world as 'Dolishan'.

We used to feed them and after a while they grew up and had their feathers and after a while they would fly away and they would come back maybe once or twice and then they would disappear. Seagulls were good pets but they were messy, very messy. Tried a cuckoo once but she was very bad, biting you all the time so we threw it away. You would never harm them … we would never harm the seagulls, never shoot a seagull.

He loved them as a boy and still does. He has been widowed recently and the gulls that still come to his house carry a memory for him of the life he shared for fifty years.

> There is still plenty of seagulls. They used to come into the house, she used to feed them a lot – I still do. I give them something everyday, they still come around, they nest in the bushes here. I throw bits of fish to them and they take it away and they feed their young with it. Now they are very territorial and if any other strange birds come along they chase it away. They come with their young. After maybe a month, they chase them away, find your own territory. They clear them off and that's the last you will see of them. They pair for life the sea birds, the sea gulls. They are always the same pair, year in year out, year in year out.

Traditional societies have a long history of empathy with the wild animals that surround them. Stillborn children were buried on the wings of whooper swans in Mesolithic Denmark. At Isbister in Orkney, more than 600 sea eagle bones were mixed in with the human remains laid in a Bronze Age tomb. In Shetland and Foula shepherds used to feed titbits to the great skuas, the bonxies, that protected their lambs from the sea eagles and ravens. And so who should be surprised that a Hebridean fisherman can see lifelong love in the gulls at his kitchen door?

5

Guillemot

The first time I found the skull of a guillemot on the Shiants, it was lying on the grey, sea-rounded cobbles of the stony beach between two of the islands. The rest of the bird was missing but the head, picked at and cleansed by the sea, was lying there perfect and complete, a diagram of life-choices, as purposeful in every detail as a gun. Its sequence of brain-case, eye-socket and beak is designed like a fighter jet, shockingly to the point, its body-forms so uncompromisingly shaped by the conditions in which it lived that it looks like something a cari-caturist or an opera designer might have thought up.

A guillemot can dive down many hundreds of feet below the surface of the sea and these bone forms are the exaggerations of nature, witness to the extremities of demand in which the guillemot must survive. What is it but a great-beaked mask worn in the Venetian Carnival to embody the qualities on which the bird must rely: violence, terror, seek-and-find, the quick and savage raid, the need to reach and kill? Imagine it scaled up from its 4 inches to something like my level. If I were a capelin or a sprat, this skull would be 6 feet long from front to back, not dead and sculptural as it is here in my hand, but driven by an alert and uncompromising nervous system. Look at it, and you see terror made flesh.

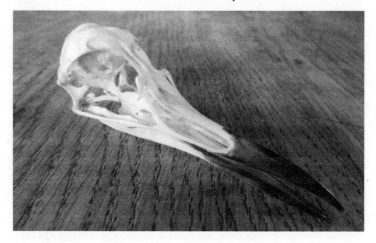

A guillemot's power-head: the encounter-zone of this brain-eye-beak
machine

In front of the brain-case, which is strongly made but not
large, about the size of the last joint of my thumb, are the vast
eye-sockets, each eye taking up as much room in the head as
the brain itself, and each of those eyes deeply protected within
an arch of bone. They are so big that the inner edges of the
eyes, when the bird is alive, very nearly touch. We do not see
birds' eyes for what they are because, unlike ours, only the
pupil is not covered in skin and feathers. But these massive
orbs are witness to the intensity of the guillemot's existence.
The only factors that limit them are the processing capabilities
of the brain, the cost to the body of running these big
information-gathering systems and of bearing their weight —
because eyes are heavy and flying with a heavy head is not easy.
These eyes are the most the guillemots can cope with, both at
speed in the air — the larger an eye the further away it can
focus on an object — and in the depths to which the guillemot
dives. If I had eyes of the same proportion, they would be two
grapefruits clamped to the front of my skull.

Compared with a land bird, the bone of the guillemot skull is not *pneumatized*, as the anatomists say. Cut open a wren or a toucan and the bones you see are not solid objects like ours. They are as light as physics will allow, in places a fine filigree that is more like bone froth or bone mousse than bone itself. The guillemot is not like that. As it is in puffins and razorbills, and in the penguins of the southern hemisphere, this bone is solid. Squeeze the guillemot skull as hard as you can, from each side or from end to end, and nothing gives. The bone is hard and heavy partly because its weight helps with diving – it is ballast – but also because this is a helmet, a power-head, the leading point in the bird's encounter with its world.

Beyond the eyes the beak is extended as if someone has taken the front of the bird's face and drawn it into the longitudinal, the bone pulled forward like gum. The beak is twice as long as the rest of the skull, the encounter-zone of this brain–eye–beak machine, the means by which the plunging, diving bird reaches into the prey. It is long for the same reason that war-swords, halberds and javelins are long: extending the radius in which victims can be caught. When Hugh MacDiarmid wanted an epigraph for a poem on finding a bird skull, a scene he decided to place on the Atlantic shore of South Uist, he famously quoted a line from an old Spanish play: '*Los muertos abren los ojos a los que viven*' – which means either 'The dead open the eyes of the living' or 'The dead open their eyes on the living.' I am not sure which of those sentences I hear when I look at the vacant eye-sockets of a guillemot.

They are divers like the other auks, the puffins and razorbills, dropping into the sea when sitting on its surface, using their wings to propel themselves downwards. I have been diving with the guillemots and they come past you in spurts, each wing-thrust jerking them forward, not the smooth momentum-fuelled movement of the seals, silkily twirling around their underwater world, but visibly hard at work, the guillemot body urged onward and downward, sometimes as they pass looking sideways at you, a sudden glimpsed recognition of the wrong sort of rubber-suited creature in their sea.

The perfect predator: a guillemot on the Shiants

It was long thought that seabirds hunted their prey by following a random path across the sea equivalent to the motion of a speck of dust floating in a room. If the fish were randomly distributed in the sea, this would be the best way of finding a shoal. But fish are not randomly distributed in the sea. As fishermen have long known, they like to linger in certain fruitful corners, or in turbulence, or at the meeting of water bodies. And that real-world distribution turns out to govern the hunting practices of the birds.

Over the last twenty years, through remote trackers and long examination of the habit of guillemots, it has appeared that the way they hunt is an impressive and resourceful combination of flexibility and responsiveness, memory and observation. The guillemot may look like an innocent, and has misled us in the past with the simplicity of its appearance: a dark, simple coat, a white chest and belly, a pointed bill, a shiny eye. It is all too easy to forget that these are cognitively complex creatures, dazzlingly successful across the whole of the northern hemisphere, exploratory, alert to changes in the world around them, and capable of knowing and remembering. Their world is dense with mobile and transient information; they have evolved to match that and use it.

No bird more than the guillemot summons the need to remember the great words of Henry Beston, the Massachusetts naturalist, whose book on the life of Cape Cod, *The Outermost House*, written in the 1920s in retreat from the horrors and despair of the Western Front, became Rachel Carson's inspiration when writing *Silent Spring* in the late 1950s. Beston was no scientist, but the attention he gave to the guillemots whose bodies he found dead but still warm on the Atlantic beach prefigures much of what the new technology is allowing scientists to understand. 'We need another and a wiser concept of animals,' Beston wrote almost a century ago, drawing on Thoreau and Whitman.

We patronize them for their incompleteness, for their tragic fate for having taken form so far below ourselves. And therein do we err. For the animal shall not be measured by man. In a world older and more complete than ours, they move finished and complete, gifted with the extension of the senses we have

lost or never attained, living by voices we shall never hear.
They are not brethren, they are not underlings: they are other
nations, caught with ourselves in the net of life and time,
fellow prisoners of the splendour and travail of the earth.

Between 2007 and 2011, scientists from the University of
Newfoundland in St John's put GPS sensors and depth-
recorders on guillemots from Gull Island east of Witless Bay
in Newfoundland (with 200,000 breeding guillemots) and
Funk Island further out in the Atlantic. The island was called
that in the sixteenth century by the first English sailors because
it stank. It was one of the last homes of the great auk, but after
that bird had been exterminated in the nineteenth century,
Funk Island became the home in summer to what is now at
least a million breeding guillemots. Retrieving the data from
the birds in these rough, wild places, Paul Regular, William
Montevecchi and April Hedd recorded over a thousand flights
and nearly 10,000 dives. Patterns became clear: guillemot
flights out from islands were long and straight, clearly directed,
not random but making for a goal, the headings matching the
return headings of previous trips and the dive sites very near
where the birds had dived before. Sometimes the guillemots
flew 10 miles or so from the island and back, occasionally up
to 40. They were going fishing where they knew there was fish
to be found, heading back out to within a mile of the place
where they had just had some success. When they got there,
they looked for the concentration of the fish ('an area-
restricted search') and the dives they then made were to pretty
much the same depths as before, usually a couple of hundred
feet down, but sometimes to 600 feet, staying under water for
more than four-and-a-half minutes.

These are learned skills. One reason that guillemots don't
breed very young but wait for six years or so before attempt-

ing to raise a chick is that they have to learn to hunt well before they can carry that burden. One way of doing that is to watch older birds, and there is some fascinating evidence that seabirds can learn from each other where to go fishing.

On the dry desert island of Isla Pescadores off the Pacific coast of Peru, the giant shoals of Pacific anchovies feed 190,000 black and white Guanay cormorants. In 2008, Henri Weimerskirch, the great French albatross expert, came to Isla Pescadores to examine the comings and goings of these birds, recording their movements every hour from dawn to dusk for nineteen days, registering the directions they took as they left and arrived back at the island, later confirming what he had seen with GPS trackers.

After leaving the colony, the Guanay cormorants circle the island and then land on the water, in a large splashy raft of other cormorants doing the same thing, between a few hundred yards and a mile offshore. They stay there for ten or twenty seconds, washing off the muck that has gathered on their feathers in the colony, before heading off to fish. Every bird Weimerskirch tracked did the same thing. There was only one raft off the island at any one time, forming an hour after sunrise and lasting until an hour before sunset. But its position, and more importantly its orientation, changed all the time, shifting from moment to moment so that it was always aligned with the stream of cormorants returning from the sea. After the birds had washed in the raft, they took off in large groups, 'one after the other, all in the same direction, forming long columns of successive groups, regularly uninterrupted up to the horizon. Interestingly, departing groups in columns flew just above the sea surface, whereas returning groups in columns flew at an altitude of 10–30 m above sea surface.'

The cormorants in the raft were probably able to detect the direction of the incoming columns 30 to 100 feet above them.

Those birds returning from a fishing trip always flew in an undeviating line straight back to the island so that their heading provided a precise bearing for the birds about to set out. This is what Weimerskirch calls a 'compass raft', conveying 'reliable social information ... updated continuously through the alignment of the raft to the largest returning columns, which are probably the most successful groups'.

Do the guillemots do this too? Another Canadian biologist, Alan Burger from the University of Victoria, spent time in the early 1980s looking carefully at the guillemots on Gull and Green islands outside Witless Bay in Newfoundland. Without exactly nailing the idea to the floor, Burger found that guillemots which had been sitting on the egg or guarding the chick for a long time – often all night when the partner was at sea, or while that partner went on long fishing trips – didn't head out to fish as soon as the partner returned to swap duties but went to sit on the sea between 100 and 700 yards offshore. There, intriguingly, they were beyond the wheel of circling birds that hangs over any seabird cliff and makes pattern-recognition difficult. Burger thought those floating guillemots were positioning themselves in what he called an 'information halo', a ring of sea just beyond the colony in which guillemots could tell the direction from which the successfully fishing birds were returning. Birds which hadn't sat so long on the ledges didn't need to bother; they could fly straight out to sea back to where they had been fishing recently. But long hours on the colony would mean that the fish had moved and the information halo was a way of telling guillemots where they had moved to. In this way, the colony inadvertently acts as a single organism, providing nutrients to its constituent parts, distributing well-being to all its members. Because each species tends to divide up the sea – shallow or deep, different sorts of prey, different distances from the colony – it is unlikely that birds of one species pay attention to the returning-direction of others. Whether other birds pay attention to the returning-direction of their own kind has not yet been established.

*   *   *

Nothing in the seabird world is more cramped than life on the guillemot shelf. Each bird lives a long time – the oldest yet caught was forty-three years old – and they start breeding when they are about five. For decades, every year, each bird returns to exactly the same tiny patch of cliff shelf, living so near its neighbours that their bodies nearly always touch. These are the smallest breeding territories of any Atlantic bird, a patch of cliff 7 inches square, over fifty occupied sites for every square yard. Since an adult guillemot is 6 inches across the chest, each breeding bird is inevitably in direct physical contact with at least two neighbours, quite literally rubbing shoulders with them. Human neighbourhood knows nothing quite like this. Guillemot life is like sleeping with the people next door, and the people next door to them, and the people on the other side, and your neighbours in the flats above and below, every summer for the whole of your life, gripingly familiar with their smells, their annoying voices, their habits of nest-building or repairing, their excellence or uselessness as parents.

The packed-in birds don't usually mind touching their neighbours. But a constant tension and torsion ripples through them. It can feel, as you look down at them, in their shuffling, jostling dark chocolate-brown mass, like panic in paralysis, the clench of boxers in a massive joint hold. A fight breaks out every twenty minutes or so. It is usually more symbolic than actual, a fight in the mind, the wave of finger-jabbing beaks

Guillemots have the smallest defended territory of any Atlantic bird

pushing and prodding at the air and the other beaks around them. There's little contact with flesh, but a billow of noise, the rasp of a high-pitched growl-shriek rolling and roaring across the colony in a stirring, trembling eruption of individual assertion, a vast, city-drowning crescendo of rage like the chant in a stadium peaking and ebbing across the multitudes.

Look at the jammed-up life of a guillemot shelf and you must read into it not the brief functionality of a thousand ants in an ant-hill, or the busyness of bees in a hive, but a densely tapestried network of longstanding relationships which have already lasted and evolved for decades, and which have been continuous over the generations for thousands of years, on this cliff since the end of the Ice Age, perhaps 8,000 years ago. They can at times look more like dates in a box or an infestation of ticks than a gathering of birds. They turn their dark-brown, protective backs to the world, facing in towards the cliff, not as a form of display but as a method of survival and protection, the gathered backs making a tortoise-shell lid against anything the world can throw at it. The feather umbrella protects eggs and chicks from predatory gulls. Guillemots will defend – and adopt – chicks that are not their own, and the sheer density with which the guillemot bodies are gathered together creates a beak-filled roof over the vulnerable chicks.

Every female guillemot lays eggs with a distinct pattern and each egg she lays resembles her others

The adult birds nibble each other, what biologists call 'allo-preening', that strange portmanteau muddle of Greek and ancient French, to mean 'the pruning of another', as if the birds were mutual gardeners, trimming and clipping themselves and their companions into the best and most lovely of shapes. Guillemot couples do this, as a way of keeping the spouse in shape and perhaps also as 'an honest of sign of quality'. If your mate can spend time making sure you are well, checking that you don't have too many ticks about you, then he must have the spare life-capacity to attend to you. A spouse that was struggling, or was exhausted, wouldn't be able to bother. Bird partners will each benefit from a spouse and co-parent in good shape; it may be that the very act of mutual grooming stimulates the production of the hormone prolactin which promotes both paternal and maternal feelings in the birds. Mutual stroking makes for better parents and better parents make for better chicks.

When many of the guillemots spend a lot of time in the colony, close up with their neighbours, they become intensely social with them, paying attention to those neighbours' feathers and parasites, sorting them out as well as their own partners, making their plumage perfect. This might be a way of reducing the stress that animals will feel if they are pushed up against each other as closely as this, of reducing the psychological costs of being in a colony. If the stress is reduced, there will be fewer fights, fewer eggs will be knocked into the sea and all members of the colony will benefit. Mutual preening of a non-mate makes for individual survival and shared happiness. It might also mean that individual birds would look after the chicks of neighbours they had come to know and maybe even like.

But there may be something else going on in the guillemot colony. In the summers of 1989 and 1990 a group of scientists on the island of Hornøya, out on the far north-eastern tip of Arctic Norway, caught eighty-five Brünnich's guillemots (the close northern cousin of the common guillemot). The birds had their territories on a set of little shelves a few hundred yards apart on a high cliff above the Barents Sea in which they fished. The scientists took blood samples from them and analysed the mitochondrial DNA of birds from the different shelves. The results confirmed what had long been suspected but never known: each shelf was the home for a different sub-family of guillemots, sharing high proportions of maternal genes between them. Not only did individual guillemot pairs come back to the shelf on which they had bred the year before, but they welcomed their daughters back to the same shelf, making room for them in a place where you might imagine there was no room at all. It looked like a matriarchy: these birds were related through their mother's genes, allowing strange males to become their partners – to avoid inbreeding – but keeping the inheritance through the female line to themselves. The guillemot habit of being nice to neighbours was in fact birds looking after their cousins and aunts, their nieces and grandchildren. The genes governing guillemot life could assume that the bird looking after its neighbour was, on that genetic level, looking after itself.

They are packed so tight not because they have to be but because they want to be. Often there are acres of unused cliff just next door to where the guillemots are making their version of Mumbai. But they know what they are doing. The best sites are consistently better at producing chicks than those on the margins.

Even within territories, the guillemots are super-conservative, laying their eggs from year to year within 2 inches of where they were laid before. And it is not entirely cooperative: good nest sites are in short supply and for the most productive sites the birds fight each other, from time to time driving long-term occupants out of the valuable properties. That ejected bird more often than not then drifts for a year or two, not breeding, waiting to claim another good spot rather than risk breeding in an unsatisfactory one. To guarantee that they won't be usurped, at least in some colonies, guillemots often stay in their territories most of the year, already back on the ledge in October after completing their late-summer moult out at sea. They may be friend-makers, babysitters, co-warriors and community defenders, but they are also profoundly at war with their neighbours.

Because of this kind of threat, and the need to ensure that eggs and chicks are never without their protectors, guillemots

are supremely loyal parents, one of each pair guarding the
young when the other is out at sea. Tim Birkhead, the great
modern champion of this bird and professor of animal behaviour at Sheffield, has described how he was once looking at a
cliff on Skomer, an island off the coast of Pembrokeshire.

> There were about twenty pairs [of guillemots] breeding on
> this particular cliff edge, some of them facing out to sea as
> they incubated their single egg. Being so close to the birds, I
> had the sense of being almost part of the colony and had
> become familiar with all their displays and calls. On one
> occasion a guillemot that was incubating suddenly stood up
> and started to give the greeting call – even though its partner
> was absent. I was puzzled by this behaviour, which seemed to
> be occurring completely out of context. I looked out to sea
> and visible, as little more than a dark blob, was a guillemot
> flying towards the colony. As I watched, the bird on the cliff
> continued to call and then, to my utter amazement, with a
> whirr of stalling wings, the incoming bird alighted beside it.
> The two birds greeted each other with delighted enthusiasm.
> I could hardly believe that the incubating bird had seen and
> recognised its partner several hundred metres away out at sea.

But there are some complexities to this story. Most other birds
are shockingly unfaithful as partners. When a coal tit lays an
egg, it is as likely to have been fertilized by her lover as by her
husband. For some coal tit wives only one in ten of her eggs
contains the husband's chick; all the others will be the product
of extramarital affairs. Seabirds tend not to be like that: they
live for decades, lay very few eggs, usually one a year, need
two parents to bring up the chicks (because food is difficult to
find, patchy and far away), and all those factors are arguments
in favour of fidelity. This is all part of the seabird's strategy for

life – breeding in colonies, late maturing, few offspring, heavy investment in each chick, commitment for the long term – as against the landbird's fecklessly short, sparkly existence, lots of chicks, plenty of local food, breeding in isolated pairs and no fidelity. Watch a pair of guillemots and you are, on the whole, looking at a pair of good, reliable citizens.

It is exceptionally easy to poke your nose into their love-life. They are open to view on their ledges. You can stare at them for days. You can put different coloured rings on their legs so that individuals can be identified. And they come back to the same spot for year after year so that long life-histories – mutual affection or lack of it, competence or incompetence as parents, what actually happens to the children, relationships with the neighbours – can all be followed, bird by bird. The guillemot is the biographer's dream and intriguingly, as bird science has revealed in the last few years, guillemots watch each other just as carefully as any bird scientist. Their uncompromisingly packed neighbourhoods provide endless opportunities for snooping, checking out the talent and calculating the main chance.

They are what is called 'socially monogamous', meaning that like human beings they are essentially monogamous but don't always manage to keep to it in practice. A key fact has emerged recently: guillemot wives call the shots. A fascinating long-term study made between 1996 and 2003 on Great Island off the coast of Newfoundland has revealed exactly what goes on in the guillemot marriage. Five Newfoundland bird scientists set up their hide next to a broad ledge on the cliffs, 5 feet by 8, with the furthest guillemot nest at the far end of the ledge and the nearest 18 inches from the biologists' examining noses. Every summer for five years, dawn to dusk, they watched sixty individually marked birds, an extraordinarily vivid psycho-theatre of seabird life in which the birds stood and laughed, fought and preened as if bred for the stage.

Some of them were undoubtedly badly behaved: one didn't feed his chick and it starved; another knocked his chick off the cliff while fighting with a neighbour; one didn't know how to incubate the egg and another simply stood next to it rather than over it, so it cooled and died. This bird was attacked by her partner when he returned to the ledge. Another guillemot decided to drive her partner off the egg, which was then eaten by a gull. Every one of these offenders was kicked out by their husband or wife during the following winter, unseen by the biologists, but evident enough the next year when the marriage was clearly over.

Sex within guillemot marriage is far more common than sex outside it, and nine times more likely to be successful, with what biologists call 'cloacal contact' occurring. This is the realm of euphemism, and because guillemot males do not possess an 'intromittent organ', penetrative rape is impossible. For any coupling to work, both birds have to be keen and so although most adulterous sex is initiated by males, most of their attempts fail. A higher proportion of couplings initiated by females are successful, with the female crouching to her lover, calling *adow, adow*, in a soft cooing invitation to her man who crows in return his guttural, gravelling cry of triumph. Recurrent failure to make it doesn't stop males attempting to have sex with females that aren't their mate. Usually they pounce on her just as she lands on the ledge, and quite often with more than one male, sometimes up to five, clambering on to her back. Sex must last at least ten seconds to work and within ten seconds the female can easily shrug them off.

So what is going on in the gender struggle on Guillemot Shelf? While the Newfoundland scientists were observing the birds, sixteen of the thirty-nine females on the shelf had sex outside marriage, one or two of them as many as twenty times a year. A good proportion, though, had sex with a bird that was not their long-term mate in the days at the very beginning of the season before the husband had turned up, almost certainly a form of insurance in case he had died over the winter and she needed a mate to replace him.

But extramarital affairs in the guillemots are more than just replacement therapy for widows or potential widows. The Newfoundland scientists established a difference between the genders. The females that had extramarital sex were the less successful breeders: their eggs had not hatched, their chicks had failed. They were looking for a superior partner to the one they had already been landed with. But the males having sex outside marriage were the most successful of breeders, the top birds, cocks of the walk. So extramarital sex was a way for an unhappy female to find a better husband. For a male, it was simply part of being the sort of top dog that neighbouring females would be interested in having sex with.

The Newfoundland scientists found that the female would have sex only with a male that had already shown some excellent paternal skills. Male birds from neighbouring nests which in previous seasons had carefully sheltered and fed their chick and which, after two weeks or so, had called the chick down from the ledge to the sea where they would look after them and feed them for many weeks: those were the top-quality birds the females would set their eyes on. Nor did they have to be next-door neighbours. Female guillemots would cross several territories to go and find an outstanding male to mate with.

When the affairs turned serious, and the bird that had chosen the new mate decided to get divorced, the partner that was dumped, called the 'victim' by the Canadian biologists, did consistently worse in later years, failing to find a new spouse or, even if one was found, proving to be worse at bringing up a chick than the 'chooser' that had gone off in search of a top-quality mate. Choosers usually thrived after making the change; victims tended to float about the shelf disconnected, condemned to isolation by the unforgiving choice its ex-partner had made.

Usually these divorces happened after the death of a neighbour that previously had been a good breeder. The situation is this: a guillemot has been putting up with an unsatisfactory husband for years. He hasn't been good with the children and he isn't much good at bringing home the fish. But no one better has been available, at least until now. Miraculously, the neighbour's wife seems to have died over the winter. He has always been a marvellous guillemot, fishing with the best of them, a perfect father. The handsome widower is now standing there alone. Obviously, any guillemot with her head screwed on is going to make a play for him. So she does. They get on famously and soon have a chick of their own to look after. The abandoned husband can shift for himself. Whose fault is it that the first marriage has come to an end?

*   *   *

In the early twenty-first century, a seabird crisis began to bite in the North Atlantic. Diminished fish stocks, shifts in the system of ocean currents and deep changes in the ocean food webs all started to batten on the birds. Not all were affected in the same way and the depth of the crisis varied in different parts of the ocean. Warning signs had already been evident for a couple of decades. It is almost certain that global warming has had a part to play, changing the temperature and salinity gradients of the ocean and setting off rippling and cascading effects all through the marine food web. Kate Ashbrook, a PhD student at Leeds University and attached to the Centre for Ecology and Hydrology in Scotland, spent the summer of 2007 looking carefully at the guillemots on the Isle of May, or at least at one very small section of them. Things were rough, as they had been for several years. The average weight of the breeding birds and of the chicks was going down from one year to the next. The fish these birds were trying to catch and eat were either not there or thin. From five o'clock in the morning until nine o'clock at night from the end of May until the beginning of July Ashbrook watched over her birds, sometimes directly, sometimes with a remote video camera. And it became clear that this shortage of food had blown the guillemot world apart.

Usually, one parent stays with the chick while the other goes off feeding. When it returns, there is a moment of greeting and meeting and they swap roles. The chick is always with at least one parent, protecting it from aggressive gulls or bad weather, sheltering it under its wing so that the insulating down, which at least for the first ten days of life provides no warmth when wet, can stay dry. In good years, with plenty of fish, the chick can spend much of its time with both parents.

The lack of fish in 2007 meant that both parents needed to be off fishing at the same time. Taking turns was not delivering the fat-rich food the chick needed. No parent meant no protection from bad weather, nor from the predatory gulls. As the season went on, and the nearby fish had been taken, the parents were increasingly absent so that by early July chicks were being left alone for almost two-thirds of the time.

Ashbrook knew this was dangerous; the gulls would surely begin to take the young guillemots. But as she sat there, three watches a day, carefully annotating the behaviour of the colony, she realized she was recording the breakdown of guillemot society. Over five weeks, she saw 465 attacks on guillemot chicks. Not one of them was by a gull; nearly all of them were by neighbouring guillemot parents that had their own chicks to protect. 'The attacks were brutal,' Ashbrook says. 'They usually involved more than one adult as chicks fled from the neighbour that attacked them first.' And the attacks were 'often prolonged and severe, with repeated jabs to the head and body'. If the chick wandered out of its own nest, that was one thing. But many of these attacks involved the neighbouring adult invading the chick's own nest. Exactly two-thirds of all the chicks that hatched were dead before they could fledge. Of these, a third died from exposure or starvation but the others were killed by other guillemots. Not for food; the bodies of the chicks were left to rot.

Some shreds and fragments of guillemot altruism survived the holocaust. About a fifth of the unattended chicks received some kind of attention from other guillemots, sheltering them and warming them against their bodies. Next to none of the guillemots that behaved kindly like this were active parents themselves, with their own chicks to look after, but were either non-breeders or failed breeders. Parents, in other words, simply saw the other chicks as rivals for their own, fit to be destroyed.

After seeing stress surging across this population of wild birds, watching the colony flailing away at its own chicks, Ashbrook suggested that a critical survival mechanism had broken down.

Most parents were away most of the time, desperately fishing for themselves and their chicks. The acts of friendship and the sense of mutuality they would have created had little chance of developing. Deprived of sociability, the birds imagined they were surrounded by their enemies and the cycle of destruction began, the mutual killing of their children, you of theirs, they of yours. In a colony that had been counted and surveyed for thirty years, Ashbrook's summer saw the lowest weight in adult guillemots, the lowest rate of parental presence with the chicks, the lowest rate of feeding of chicks, the lowest weight in chicks at every stage and, as an inevitability, the lowest-recorded survival of young birds. The neighbours – many of whom will have been their uncles, aunts, cousins and grandparents – had killed them.

*   *   *

Even though the Newfoundlanders continue to shoot between 200,000 and 400,000 guillemots every year, that uncompromising and barren rock rising out of the Atlantic, capped with dense forests and fringed with often brutal shores, draped in fogs in which, if you stayed 'on Deck for two hours you will get wet to the skin', remains one of the world headquarters of the species. Some 2.25 million can still be found there every summer and, since before the arrival of any Europeans, guillemots, preened, glossed and bright eyed, had been one of the foundations of the world view of the indigenous people of Newfoundland, the Beothuk.

They had probably arrived at about the time of Christ, an Algonquin-speaking people, paddling in birch-bark or caribou-skin canoes from Labrador across the Strait of Belle Isle. Vikings had encountered and fought them at the end of the tenth century – they called them Skraelings – but it was the Beothuk's love of an iron-rich earth pigment which meant that, when Europeans met them again at the end of the fifteenth century, they were given a name that would stick. The Beothuk liked to stain their clothes, their bodies, their weapons, their houses and their dead with ochre, and so for the visiting Europeans these people became the Red Indians.

Like many coastal people of the pre-modern world, the Beothuk organized their lives and deaths according to the seabirds. They hunted and fished using bows and arrows, spears and darts, clubs and slings, a Stone Age culture without metal, living in wigwams covered in seal skins. In summer they slung their coats over one shoulder like a hussar with the fur outward, in winter the coats were worn closely with the fur on the inside. They kept 'their haire somewhat long, but rounded ... Behind they have a great locke of haire platted with feathers, like a hawke's lure w[i]th a feather in it standing uprighte by the crowne of the head.' Early English travellers took musicians, Morris dancers, buttons, beads and ribbons with which to enchant them, but the relationship soon fell into one of mutual suspicion, theft, murder and revenge. By the early nineteenth century, the Europeans had exterminated them. No Beothuk lived after 1829.

Nevertheless, enough of their thought-world survives for the Beothuk to provide a window into the deep past of our relationship with seabirds. For them, the guillemots were realm-shifters par excellence, at home in all elements, solid, liquid and aerial, the great appearers and disappearers of the natural cycle of the years, the visible and living connectors of an integrated universe.

Even now, Newfoundland is teeming with seabirds. Every summer, Beothuk men and women canoed out to Funk island, across 25 miles of open ocean, to gather the protein, almost certainly using the birds themselves as guides through the often foggy waters. Many of their stone arrow-points have been found there. There might have been, at most, about 1,000 Beothuk people. The seabirds that gather in Newfoundland every year weigh 52,000 tons. It was a system that could last.

Just as Icelanders still smoke the birds – or nowadays mostly freeze them – to make them last into the winter, the Beothuk boiled the birds and dried the flesh, hard-boiled the eggs in birch-bark pots, dried the yolks in the sun, making a powder of them, before mixing them with fat and stuffing the resultant nutritious paste into bags made of animal intestines. What the summer offered, the ingenuity and technology of man transformed into winter food.

Much of the summer, they spent in or near seabird islands, where all but 5 per cent of the guillemot colonies are found, away from land predators. As the year rolled by, so did their hunting practices: in early summer, the collecting of eggs; in high summer, the taking of birds and full-grown, fat-rich chicks; in the autumn, the catching of migratory birds and the salmon running in the rivers and the hunting of caribou and beaver; in late winter, some reliance on ptarmigan and grouse, and migrating harp seals, before the hungry months of early spring and the longing for the return of the magic creatures of summer. Wooden storage pots, bows with gut strings, arrows with flaked stone heads, cooking skills and canoes were the key technological links that could make this a viable life. Once European fishermen started plundering the bird colonies and shooting at the Beothuk egg-collecting parties in their canoes, the Beothuk were en route to extinction, forced to attempt life inland where the summer resources were not rich enough for them to last the winter. They needed seabirds to survive, not over the wonderful summer months out in the camps on the islands, but in the winters that followed. The last of them died of disease, weakened by malnutrition.

But even when they were forced inland by the Europeans, the Beothuk continued to bury their dead by the sea, usually accompanied in the grave by seabirds and by objects designed and carved to represent the souls of the birds. The birds, as the archaeologist Todd Kristensen says, 'served as spiritual messengers who conveyed souls of the dead to an island afterlife'.

One of the most graphic of all these burials was found in September 1886, on an island in Pilley's Tickle, a deep, wooded inlet on the frayed north-eastern coast of Newfoundland. At the end of winter, icebergs sometimes drift and ground in the tickle, blocking the routes for the ferries. In the autumn of 1886 a family was there collecting berries. On a shelf of rock under a cliff, a boy found the ground giving way beneath him, his feet breaking down through sheets of birch bark and a canopy of arched sticks into a fourteenth-century Beothuk grave. Inside was the body of a child, a boy about ten or twelve years old, lying on his left side, wrapped in a beaver skin shroud, the flesh side turned out and smeared with red ochre. Inside, he was dressed carefully in skin trousers, neatly sewn together, and fringed at the sides. On his feet he wore moccasins, also fringed, with a couple of spare pairs of the same pattern beside him, all of them sewn with fine deer-sinew stitches.

Arranged around the body of this much-loved boy was a collection of treasures and companions for life-in-death: a small wooden image of a man, two miniature birch-bark canoes, their hulls with deep V-shaped seagoing profiles, miniature bows and arrows, paddles, a couple of small packages of red ochre tied up neatly in birch bark, a package of dried or smoked fish, salmon or trout, also made up in a neat parcel of bark and fastened with a network of rootlets like a rough basket, and most important of all, carved bone pendants and the feet of some guillemots, sewn to the shroud and smeared with red ochre.

His people had equipped him for the journey into the next life, to the land of the dead, described by the last of the Beothuk in the nineteenth century as an island in the country of the good spirit where they could hunt and fish and feast. Beothuk heaven was a seabird island. Heads of seabirds accompany several Beothuk burials. In one of them a few seabird skulls were found together in a 'medicine bag'. The ornaments were there to help the dead on their way. They were usually made from caribou bone, split and ground into narrow flat plaques, before carving and coating in ochre. More than 400 survive, collected from graves all around the Newfoundland shore.

These pendants are, in effect, the power of birds transformed into wearable objects. It is bird-power jewellery, not like our jewellery enshrining the everlastingness of gold and diamond, loving its resistance to change, but the opposite, objects formed by entrancement with the life of seabirds, an attempt to capture their beauty, intelligence, range and meaning, in the form of long, asymmetric primary wing feathers, three-toed seabird feet, or long forked tails like the tails of the Arctic terns that breed in Newfoundland.

All over the Americas and Eurasia, there is a close connection between birds and the shaman's ability to shrug off the physical world. A gold shaman's pendant from Colombia now in the British Museum shows the man himself turning into a bird as the narcotics take hold and he begins to fly across space and time. An enormous grin spreads across his face as his wings open and his feathered tail flares behind him. Among the Evenk of eastern Siberia, as Kristensen and Donald Holly point out, eagles and cuckoos were messengers to the upper world, cormorants and divers to the world below.

What the traditional societies saw in the life of the birds translated into their significance beyond the physical. Seabirds were in a category of their own, powerful in all three dimensions. The forms of the bone jewellery made by the Beothuk are a reflection of that comprehensive power. The seabirds could hunt in water, diving for the depths, using their webbed feet; they could hunt in air, using their wings; and they could travel the world, making their way across the widths of the ocean, to an extent surely unguessed at by the Beothuk but clear enough from their appearance and disappearance with the turning of the years.

The Beothuks were buried on seabird islands with seagoing equipment and seabird jewellery, even when the Europeans had driven them from those shores in life, because the seabirds were for them transcendent, climbing beyond earthbound categories. No geography could be more sacred. For Kristensen and Holly, the seabird cosmology emerges from the facts of Beothuk lives:

> During late spring and summer the Beothuk visited seabird colonies to obtain meat and eggs. Depending on skill and weather these expeditions surely involved varying degrees of difficulty and danger. They involved scaling cliff-faces, landing fragile canoes on rocky shores, and navigating Newfoundland's often foggy shorelines. In particular, 60 km voyages to Funk Island would have entailed great risk and required navigational acumen. But once there the Beothuk encountered an island teeming with birds and covered with eggs. It was perhaps just the sort of trip and destination that provided ideological inspiration.
>
> The island rookeries were places where birds came to give birth to new life before departing for distant points beyond the horizon – but perhaps not beyond the imagination. We envision that the Beothuk hitched the souls of their dead to this great mystery of bird migration; they buried their dead in places where birds created life (islands) and hoped that when the birds departed they took the souls of their dead with them to their island afterworld.

It is a shared inheritance. Wherever people have lived with and on the birds they have come to regard them in this way, as companions and guides. On the Bering Sea coast of Alaska the Yup'ik, an Inuit people, with cousins of the same ethnicity still in Siberia, have survived to tell the stories which for the Beothuk of Newfoundland can only be guessed at. They spend their lives quite literally clothed in the seabirds. Underneath their waterproof outer clothing, of fur and gut, they often wear parkas made of seabird skins, not only because of the warmth this next-to-the-skin air-trapping fabric gives them but because birdness, if such a word could exist, is what the Yup'ik hunter has to strive to achieve. Few are made now but multiple examples still exist, of auklet and puffin, eider and guillemot. Most are fragile, worn in the daytime with the feathers turned inwards to protect them from abrasion, used as the most precious of blankets at night with the feathers on the outside. Every year the returning birds provided skins for new parkas.

For the seal-hunting Yup'ik, this closeness to the birds cannot be effective if superficial. It must run deep. No Yup'ik man should eat the marrow of the wingbones of a goose because if he does he can only ever paddle slowly when at sea. Nor is the good hunter, out in his kayak on the ocean, pursuing the seals, to appear 'like a seabird'. Likeness is not enough. The Yup'ik when hunting wore bentwood hats with long peaks, decorated with white clay and black paint, to hide them from the seals as they crept towards them. The peaks would keep the sea spray from the hunter's face when the waves were breaking over him, but more than that they would embed him in the ocean world, allowing him, or in some stories both him and his kayak, his whole presence on the sea, to become a bird.

For a Yup'ik to be a good hunter, he needs to be thought a seabird in the seal's eyes. It is the Yup'ik assumption that the good hunter actually becomes a seabird, in many stories disappearing behind an ice-floe, only for a bird to come tentatively out on the other side. This is no part of the urge to mastery. The hunter does not win the seal by dominating it but by such a total immersion in the identity of the bird that his own presence melts away and into it, their *Umwelten* merging in the ice-scattered sea.

6

Cormorant and Shag

The most sinister seabird in English literature makes his brief and unforgettable appearance in Book IV of *Paradise Lost*. Satan, the great fallen Angel, the broken prince of Heaven, is in a turmoil of doubt and despair. He knows he must wreak revenge on God by corrupting man, God's most precious creation, but the prospect troubles him. Faced with the task, 'his grievd look he fixes sad', but for all his resolve he cannot escape Hell, because 'my self am Hell', and his conscience 'boiles in his tumultuous breast'.

Eden lies waiting for him somewhere far below and Milton, in the darkness of his own blindness, imagines the great spirit flying down through space to the 'soft downie Bank damaskt with flours' on which Adam and Eve both lie. The flowering trees of Paradise fill the air with their scents, and the breeze blows them out into the spaces of the universe. As Satan slows on his voyage, relishing the winds from that perfumed world for which he lusts and which he must destroy, the idea of a jasmine-flavoured cosmos suggests to Milton stories he has heard of European sailors venturing into the promise-laden air of the Indian ocean:

As when, to them who saile
Beyond the *Cape of Hope*, and now are past
Mozambic, off at Sea, North-East windes blow
Sabean Odours from the spicie shoare
Of *Arabie* the blest, with such delay
Well pleas'd, they slack thir course, and many a League
Chear'd with the grateful smell old Ocean smiles.

Satan is from the other end of being, troubled and shadowed, inwardly complex, dark, embroiled with death, and as he approaches he sees the bosky beauty of Eden below him:

Groves whose rich Trees wept odorous Gumms and Balme,
Others whose fruit burnisht with Golden Rinde.

Then Milton springs the seabird moment for which this has been the preparation. In Eden, among these flowering riches, was 'the Tree of Life,/High eminent, blooming Ambrosial', from which Adam and Eve could pick the 'Nectarine Fruits', grazing on the sweetness and perfection of the world God had made for them. All is colour and light and all of it the stage for Satan to overleap the walls, when:

Up he flew, and on the Tree of Life,
The middle Tree and highest there that grew,
Sat like a *Cormorant*; yet not true Life
Thereby regaind, but sat devising Death
To them who liv'd.

Milton wasn't making this up. Aristophanes, Plutarch, Chaucer, Erasmus, Shakespeare: everyone has given this bird his evil part, and for all of them the greed of the cormorant shag was

The bird who 'sat devising Death/To them who liv'd'

allied with power and prerogative, the assumption by the powerful that they should consume what others had.

Across the whole of Eurasian culture, the cormorant has been the bird of greed, the enemy, the dark one, the usurer – 'Shylock' is a rough transliteration of the Hebrew name for it – the water-crow, eel-rook, fish-hawk or, in China, the thief and black devil. It has always been fused in people's imaginations with its near-cousin the shag, as dark satanic birds, the sea ravens which come from some other deeper, fiercer and more dangerous world.

When I first saw the shags on the Shiants and spent days with them as a young man, looking to see what sort of animal they might be, I understood them, I now realize, just in the way Milton had portrayed his beloved arch-devil. They were creatures from another realm, whose whole strange, clumsy, threatening manner is more distant to us than any guillemot or puffin could ever be. I remember climbing up into one of their colonies high on the western slope of the islands. As I came over a lip of rock, there was the shag right in front of my face, a foot away, juddering and hissing, its whole head shaking in rage and fear, terrifying as much as it was terrified of me, a fluster of beautiful dark green iridescent feathers in the mayhem of kelp stalk and guano that was its nest. Ancientness bellowed at me from inside the filth-lined crevice, where, in the shadows, two or three featherless, scrotal-skinned shag chicks writhed like embryo sea monsters from the past. This was all depth. Nothing neat or sweet, but powerfully attractive and seductive, satanic even, foreign, alluring, with a glimmering nightclub glamour about it, as if I had come on some transitional creature, half-pterosaur, half-bird, gleaming in the oily dark of its feather sheath, blazing in the deep rich yellow of its gape and gizzard. Around me, in the air, the other shags coming into the colony honked and hooted, the deep, guttural cry of the arriving males, something from beyond any world I had ever known.

That idea of the shag is pervasive in Western minds. Fishermen have always shot the cormorants, seeing them as the great enemy of fish and hence, perhaps, of mankind. William D'Urban, the Victorian naturalist and author of *The Birds of Devon*, was particularly impressed by the beauty and glowering, demonic heroism of the shag.

> The adult bird in its dark green plumage in the early spring, and with its conspicuous crest, is very handsome, although, to our eye, there is something 'uncanny' in the appearance

both of the Shag and the Cormorant, as Milton doubtless thought when he selected the Cormorant as the bird whose form he represents the great Enemy of mankind as assuming.

The Shag is an extremely wary bird, and does not often become the spoil of the gunner, and its compact feathers are almost like a coat of mail. An Instow boatman, who in by-gone years was our companion when we went after Duck, had many marvellous tales to tell of this invulnerable bird. One day he got near one as it was sitting asleep and drying its feathers in the sun, after the manner of Shags, on a sandbank. He shot it, picked it up, and tossed it into his boat, and a short time afterwards landed to crawl over the mud with his long duck-gun after a flock of Wigeon. When he was some distance away he turned round and looked towards his boat, and there, he declared, was his supposed dead Shag seated on the gunwale, shaking the shot out of its feathers, so that he could hear them rattling on the water, and the next thing he saw was the Shag flying cheerfully away!

Mysteriously, there is some kind of evolutionary reality to the long and shared perception of the shags and cormorants coming from another level of creation. They are certainly among the least evolved of the seabirds, nearest in body-form and perhaps in lifestyle to the first fossil seabirds that emerged about 100 million years ago. They have not adopted the full-ocean life which later-emerging birds perfected. They do not travel thousands of miles out into the riches of the ocean but stay near to shore, diving down with their giant, webbed feet to the seabed where they catch the bottom-lurking fish. Unlike the auks or the fulmars and petrels, the shags lay many eggs, usually three, although I have seen six in a nest, white, simple-looking things, not the decorated beauties laid by a guillemot or an oystercatcher. They do not have the warming, featherless brood patches other birds have on the underside of their bodies, but must place the eggs on the webs of skin between their toes, shuffling into the nest when taking over from their partner. This idea is not acceptable in evolutionary terms but it is difficult not to think of the shag as a slightly Heath-Robinsonish, exotic and unrefined attempt at seabird design, as if it were an early car, full of flamboyance and inefficiency, florid and demonstrative, set against the sleek, modern, self-containment of, say, a guillemot, which is as neat as an iPhone.

I have been out ringing shags with a team of birders who were taking them from their nests using short poles with metal hooks attached to the end. The hooks encircled the shags' long sinuous necks and pulled them from the nests, so that rings could be attached to the legs. Usually nowadays a metal ring, to be recovered on death or recapture, is added to a coloured plastic ring which can be read through binoculars. And although there is all kinds of scientific justification for doing that to a bird, allowing its life- and death-habits to be measured and guessed at, I watched the ringers at work and felt distant from them. It seemed to me that you could hold more of the shag, see more of it and know more of it when standing looking at them in flight than when grabbing or clutching at them, one arm around the chest and body, the other firmly clamped around the lacerating bill. As soon as a shag is in a man's hands, its *Umwelt* has gone. That whole imaginative, intense, life-encompassing and evolved halo evaporates when the bird is held. They are diminished creatures in the hand, their aura of power and strangeness lost, gripped by terror more than anything else, confused and hopeless in a world of nets and pliers, their wings sometimes caught in the mesh of nets like awkward elbows, their fluster a collapse of being.

I hold the young shag, which is just on the point of emerging into adult life. Its dark-grey fuzz of down mounds up between my own anxiously gripping fingers. I have its whole body encircled in the ring of my two hands. Its neck, twisting and turning, is somehow serpentine for me so that its distance from the reptiles seems particularly short. With its black eye, and the hooked grey bill, already with a rim of pale yellow on its lips, it reaches round towards me, looking for the animal that is threatening its existence. Its gape suddenly opens as wide as its whole head is long, and I look down into that grey-yellow gullet, the bright anxious eyes above it yawing and shuddering at the enemy. The bird is undeniably here in my hand, warm on my skin, wriggling for escape – but with what strange loss of presence, as if I have already squeezed its essence out of it. The ring I help put on its leg is a shackle, not on the body – because the hard, lightweight metal is designed to be unnoticed – but on the idea of the bird.

No ringer had the slightest intention of harming the shags – the very opposite – but when I saw one of the men come back from his day among the boulders, the whole of his chin and lower lip scabbed in the blood that a shag had torn from it, the outer hook on the scaly bill capable of slicing through far more than a ringer's flesh, I felt mysteriously relieved at the act of revenge.

There is another way of meeting these birds. In China, people have been living with cormorants for centuries as close companions. They are bred in captivity and the eggs brooded by chickens, because the mother cormorants are not that attentive. The dark and naked chicks emerge after a month's incubation, when they are transferred to cotton-lined baskets kept in a warm room. The young cormorants are wrapped in cotton wool and fed with small morsels of tofu and pellets of raw fish, preferably eel.

In a way known to Chinese fishermen at least since the seventh century AD but understood by Western science only through the experiments with geese made by Konrad Lorenz in 1930s, the young cormorants 'imprint' on their owners and nurturers. The late nineteenth-century French dilettante huntsman Pierre-Amédée Pichot tried raising a few himself:

> This web-footed marine bird is very easily tamed. His heart is very near to his stomach, and one may be reached by way of the other ... If you feed him out of your hand, you will have trouble to prevent him from following you everywhere, ascending the stairs behind you, perching on your furniture, and leaving on all pieces incontestable traces of his rapid and abundant digestion.

Each cormorant on a Chinese fishing boat knows and returns to his own place on the gunwale

But the Chinese had long since perfected the art. There was no need to keep a trained cormorant tethered; the birds would return from their fishing expeditions and disgorge what they had caught with no more encouragement than a good feed of fish afterwards. A narrow ring of hemp and straw was tied around their necks and attached to a harness across the back to prevent them swallowing the fish entirely but the best trained did not even need the ring and would bring the fish back unasked. The fishermen bred them like prize spaniels, focusing on variations in the plumage, with many chequered or even pure white individuals, and they became as docile and obedient as dogs, knowing their master and his boat, returning always to the same spot on the gunwale which belonged to them, understanding his commands and knowing the places where they had caught fish and were likely to find them again. One fisherman might have a flock of ten or twelve birds, but would not rush all of them into the water at once, encouraging only one or two to dive at a time, calling them by their individual names. When he saw they were tired, he took them in and fed them, while sending another pair to go off fishing, sometimes banging the water with a long bamboo pole to urge them them on. When they returned they would jump on to that pole so that he could bring them into the boat with their catch.

This daily and humane closeness revealed an aspect of cormorants which modern science is only now coming to recognize:

There is a good deal of individual character in the birds. Some are intelligent, alive, and alert; others are dull, lazy, and sulky. Some are more expert at diving and catching than others. European writers have described the cormorant as 'solemn, weird, uncanny'; but such words are elicited by impressions of our mind, and are not objective characteristics.

For a brief moment in the seventeenth century, cormorants were brought by the Dutch from China and kept for sport in France and England as diversions for the monarchs, who liked to watch them fish on ponds and rivers at Fontainebleau or where the court was out hunting. James I even paid a Master of the Royal Cormorants, which were kept in enclosures in St James's Park in Westminster and taken 'hoodwinked' – under blindfold – from there to the Thames where they fished. Milton as a sighted young man would have seen them in London. When writing *Paradise Lost*, already blind, he was living in Petty France, just south of the Park, from where he would have heard them hooting and calling. And from there they flew on into his memory: the blind poet thinking of the dark-skinned, hoodwinked bird, reimagined, while he composed at night, as the Prince of Darkness alighting on his journey from Hell in the lost world of bright perfection.

The time-gap between man and seabird – the strangeness which goes back 300 million years to the moment we had a common ancestor – is not as deep a disconnection as, say, between us and the fish or us and the insects. It is a chasm of time, but when you look into a shag's eye, its blazing green in the dark and glimmer of its head, a little bead of intelligence regards you from some distant province of creation. There is a kind of understanding there, a mutuality you could never feel with a wasp or bacteria. Frans de Waal has asked recently, 'Are we smart enough to know how smart animals are?' That is the question asked by the shag's eye.

Look at him on his springtime nest. His dignity is diminished a little by the tufts of fluff that hang around his bill. He has been attending to the nest, perfecting it, and one or two pieces of down have stuck to him. One shag, not far away, has decorated hers with tendrils of sea campion, strung around the rim like a bridesmaid's coronet. When both parents are away for a moment, their neighbours will always steal particularly attractive pieces of string or seaweed. Early in the season, complete nests are within a few minutes frequently removed by neighbouring birds when the owners are temporarily away. Scarcely a minute passes when they are not checking and tweaking at the rim, ensuring the cup in which the eggs are laid remains whole. But this is more than engineering: like the puffin's bill, or the guillemot's assiduous preening of his spouse, a handsome nest is a signal of a shag's excellence.

His body is his blazon. His head wears his tall forward-arching breeding-season quiff. He ripples in the snakeskin robe of his plumage, the edge of each feather darker than the rest so that his green back is netted in a darker green which shimmers blue in some lights, black in others, like the heart of a peacock feather-eye. The Chinese, seeing these markings, called him 'the black-headed net', but this is another of the roles played by melanin in the plumage of seabirds. The substance that makes the feathers dark is also a strengthener, and so the shag's feathers are darker where they need to be stronger, on the edges where life is most likely to abrade them.

Watch a pair of these ferocious, glamorous birds together on the nest and you will witness a slow and careful ballet of tenderness and sweetness unfolding between them. It looks deeply sexual, each shag reaching in an exploratory way around the other's neck and head, a mutual exploring of the other body and its geography. 'Licence my roving hands, and let them go,' Donne wrote in his famous poem, 'Before, behind, between, above, below.' That is the sensation of two shags pair-bonding at their nest. This bird is not simply a predator, as these creatures are so often called, as though an animal were merely its economic role in the structure of life. People see shags like that because their fearsomely hooked bills are so hard. But watch two shags together and they do not meet each other only with those high, hard points of themselves, their armoury. They meet whole body to whole body, each neck curled and turned around the other. There are no soft palps to their fingers but their breasts rub, as if communicating intimacy there, with the fullest part of themselves. When the bill does come into play, and one bird pays attention with it to the other's neck, it nibbles, just lifting the feathers so that afterwards, like a love bite, you can see where the birds have told each other that they are a pair and belong together.

A few lifted feathers show where the birds have been nibbling at each others' necks

For all that, though, the cormorants and shags are not the most loyal of partners. Some land-based waterbirds, including the flamingos and the herons, effectively divorce their partner every year. No heron or flamingo marriage lasts more than a single season and marriage is probably the wrong word for it. At the other end of the scale, the divorce rate of the wandering and waved albatrosses is zero: they remain faithful to their partner for the whole of their immense lives. Other albatrosses are not far short, the sooty albatross divorcing on average once in twenty years, the black-browed only slightly more often. The fulmar almost never divorces and the puffins vary from 7 to 13 per cent a year. Only the guillemot of the true ocean birds, constantly on the lookout for a better option in its crowded and busily self-referential colonies, falls away from the loyalty ideal, divorcing at a rate of nearly one in five marriages every year.

The shags and cormorants are intermediate, varying from inevitable annual divorce in the Crozet shag, one of the blue-eyed shags of the Southern Ocean, to 88 per cent in the flightless cormorant, 58 per cent in the Antarctic shag and 36 per cent in the European shag. The shags are only partly committed to the monogamous life, loyal enough when married but rarely married for long.

There seems to be a connection in birds between lifestyle and the size of the brain. Along with his co-researchers, Nathan Emery, a bird cognition specialist at Queen Mary, London University, has established a rule of thumb relating brain-size to a bird's lifestyle. Intriguingly, the more faithful a bird is to a mate, the more likely it is to have a relatively large brain. Emery looked at 480 different species. If they stayed with a single mate for the course of a year, their brains were just below average size in relation to their bodies. If they were faithful for more than a year, but not for the long term, that brain-weight began to nudge above the average. If they were long-term mates, like the puffins or the fulmars, brains were well over average size, up to 15 per cent bigger than the mean. Promiscuous birds were at the average. Those in which the male had many wives were well below it. The birds with the smallest brains were those in which the females were completely promiscuous. Radically unfaithful females had brains that were 20 per cent or more smaller than the average.

Birds don't live in complicated, multi-level societies as mammals tend to. Bird colonies are simple gatherings of couples, with an outer ring of adolescents looking on. There is no government in a bird colony, and so birds don't need the kind of big brains mammals require to conduct their subtle social and political lives. Why then do bird brains tend to be so large in relation to their bodies? Emery's theory, based on the evidence of big brains in birds with longstanding relationships, is that birds need to be clever to understand their partner. If they can get in tune with another bird, understand its foibles and predict its behaviour, that will give them an edge on all other birds who can't form such deep and intricately inter-locked partnerships.

What you are witnessing, as the shags nibble carefully at each other's necks, as they make good the nest for each other, as they honk and hiss on arrival and departure, as they rub themselves into intimacy, is evidence of the tight bonding between birds which later-evolving seabirds elevated into a principle of life and survival. Love matters for seabirds, because a harsh environment, in which food is difficult to find, can make raising healthy offspring more difficult. Long-term monogamy will be an advantage in these conditions as both birds can attend to the well-being of the chick, providing a double support system which infidelity or promiscuity cannot match.

The sequence is this: in a difficult environment, better-looked-after chicks survive; monogamy creates a better start for them but also requires some subtle cognitive skills in the parents, each recognizing the other's moods and needs, and using that information to predict their behaviour; you have to be clever to be a good spouse and only the progeny of good spouses will survive. A difficult and shifting ocean will in the end educate the genes of which the birds are made into long-term monogamy.

Seabird chicks that are going to learn these skills must be born capable of learning, and it may be that the long incubation time of most seabirds, approaching two months in many, longer in others, and the long feeding of the chick in the nest, sometimes several months, is required to make the necessary neural networks in the young bird's brain without which it will not, over the coming years, be able to understand its world. It is a long and difficult business, which is probably why seabirds do not mate until they are several years old. Maturity is required if the parents are to catch enough fish for their chick, but also to manage a good relationship with their mate. As Nathan Emery and his co-authors have written:

> keeping track of the accumulating, subtle behavioural characteristics of a bonded partner over the course of a relationship requires ... a form of relationship intelligence, which enables them to read the social signals of their partner accurately, respond appropriately to them, thereby predicting their behaviour and hence resulting in the stability of a successful partnership with mutual benefits for both parties.

It is possible that the extraordinarily docile and compliant behaviour of the Chinese fishing-cormorants is an extension, similar to a dog's, of a natural aptitude in the bird to understand and attend to the needs of its partner, whether that partner is cormorant or man.

In the recent past, it would have seemed absurd to talk about birds in this way, but a recent study by Seweryn Olkowicz from Charles University in Prague and his colleagues has revolutionized the question of just how birds with their tiny heads are able to achieve their feats of understanding and sociability. Olkowicz has found that birds have twice as many nerve cells in their brains as mammals of a similar brain-size. Bird brains are crammed with information-transmitting cells, smaller than those in mammals and with fewer connections, but packed with processing power, miniaturized for flight.

A shag, poignantly enough, is only part-way clever. The cleverest birds are the corvids, the crows and ravens, and the parrots. A shag cannot, as for example the spectacled parrotlets can, give each of its chicks a separate name. Nor can it imagine what another bird might have understood, or make tools to get at food, as many crows can. Nor does a shag quite fit the brainy pattern of long-term fidelity, as it changes partners every three years or so. It doesn't focus on the single precious chick but has a way of laying too many eggs. The ancient shags, the earliest maybe 70 million years ago, were probably freshwater birds, which only later migrated to the coast, and there it has no more than dipped its toe in the sea world, half responding to the oceanic disciplines. It rarely dives very deep, almost always less than the auks for example, and that has resulted in its most characteristic stance, on the shoreside rocks after diving for fish, standing drying with its wings spread like the dark angel of a fallen world. That look must be one of the reasons Milton chose the cormorant as his

Satan bird. No other seabird, apart from the cormorant's relative the darter, needs to dry its wings in that way and the
reason it does is intimately connected with its shallow-fishing
habits just offshore.

Marine mammals can rely on blubber to insulate themselves
against the cold of the sea, but blubber is heavy for a bird, and
waterbirds usually wrap themselves in air trapped in their
feathers to keep warm. Trapped air, when diving for fish, is
highly buoyant and birds have to struggle to get or stay down.
Small diving ducks, for example, most of which fish in the
shallow upper few feet of water, are said to put up to 95 per
cent of their mechanical work to counter buoyancy during
dives. One answer is to dive deep. Water pressure squeezes
the air that remains lodged between the feathers so that down
at its fishing depths the bird does not have to fight its own
buoyancy. Some diving birds are known to breathe out before
they dive, reducing the oxygen in the lungs, and others have
muscles in their body which pull their chest- and back-feathers
tight to the body as they dive, to increase their density.

Shallow-divers like the shags drive themselves to the sea
floor entirely with their feet. As they sit there gleaming in their
fortresses, this one part of them does look absurd: flapping
finned galoshes, the webs stretched between all four toes, like
a collapsed umbrella on the end of each foot, slapping against
the rocks as they make their way around the colony. Their
wings are far too large to work well under water, creating as
much backward drag on each stroke as any forward impulse,
and so these feet are central to the need to feed. But no great
depth means no water pressure to squeeze the air out of the
feathers. And so the shags might have been faced with a big
buoyancy problem.

To counteract this, they have developed a particular kind of
feather, quite unlike those of other birds, with a much more

open structure than most, at least at the outer edges. The shags need to achieve a working compromise: not too buoyant but not too sodden, as the metabolic costs of keeping warm would be difficult to sustain if it was constantly soaked to the skin. So the outer parts of shag and cormorant feathers are very open in structure, allowing water to soak into them and expelling the air as soon as they are wet. But the inner part of each feather is much more dense, resists water penetration and remains full of air, so that that the bird can keep warm. The balance is finely calibrated. Cormorants which, however reluctantly, were made to swallow recording thermometers have shown that their body-temperatures, as they dive to 50 or 60 feet and swallow large lumps of cold and chilling fish, remain almost constant. The only cost is when they return to land and need to dry the outer edges of the feathers to restore their insulating properties.

Learning how to survive is exceptionally difficult for a shag and the mortality among the young birds can vary enormously. The chicks are born strange and naked, develop a brownish fluff after a week or two, and stay in the nest for about eight weeks before leaving for the sea near the colony. Like their parents they roost ashore each night, and they continue to be fed by them after they have left the nest.

Most small landbirds live, on average, for a little over twelve months. Most deaths are among the very young. Of a brood of baby thrushes that have just left the nest, 14 per cent will die every week. Over 90 per cent of all chicks will die in the first year. But just over half of all adult robins will also die each year so that more than one in every two birds you see in your garden will be not be there in a year's time. The treasured robin on the gatepost or thrush on the apple tree is likely not to be the one you saw last year, but its successor and replacement.

These landbirds are on a quick-return, go-for-it gamble: if each pair can lay many eggs, often in more than one clutch, and raise many chicks, the chance is that one will survive to join the breeding population in the following year. They are in an attritional relationship with their environment, the breeders like commanders on the Western Front flooding it with sufficient raw recruits for them to hope and even expect that one or two will make it through. Most seabirds take the other approach: careful, low risk, low reward, slowly and persistently feeding single spies up the snaking trenches of their battleground, assuming that only the strong, the clever and the well-equipped will have any chance in the enveloping hostilities of the ocean.

The shags and cormorants hang, as in so many other ways, in the middle ground between these poles, suggesting that their life-habits may represent the residue of what they learned tens of millions of years ago when their ancestors were landbirds and they were first inching towards the water. In May 2003, Francis Daunt and his co-researchers attached miniature activity-loggers to the four- and five-week-old shag chicks from eighty nests on the Isle of May, off the east coast of Scotland. The young birds fledged when they were about seven weeks old and went out into the sea, gradually spending more and more time hunting for fish until forty days later they were fishing for up to ten hours a day, far longer than the adult birds.

Daunt understood that the young birds were learning to fend for themselves, but in those first weeks after leaving the nest the parents continued to look after and feed them. By the shortened days of early November, the young shags, which had still not learned to fish well, did not have enough time in the day to catch the fish they needed to survive. The competent adult shags were still able to rest for a few hours of each day but the younger generation were unable to spend any daylight time not fishing.

By mid-December, the starving young birds began to die, their colour-ringed bodies picked up on the shores of the island, the mortality peaking in January and February, when the proportion of young shags dying was five times that of the adults. As the days started to lengthen again, the moribund birds were too weak to take advantage of the daylight and, for two or three weeks before they died, the amount of time they spent fishing gradually declined until they were looking for food no more than three hours a day, too exhausted to assuage the hunger that was killing them.

Each year there is a devastation of the young. When three years later in the spring of 2006, Daunt and his co-researchers looked for the survivors of the birds they had ringed in 2003, they found 8.9 per cent of them had returned as adults. That was a higher percentage than usual. The proportion of the general shag population which survives three years to return to the Isle of May is 7.4 per cent. Over 92 per cent of all young shags die before then.

'From Cradle to Early Grave', Francis Daunt called his paper describing this research. It is a phrase which lends a kind of seriousness to one's vision of the shag, so that whenever you see a male bird at his nest honking the deep *ark, ark, ark* of his primordial song, or come on a female head-shaking and hissing at you with an urgency that sounds like fat splashed in the pan, that is a moment to consider the hazards it has passed, the enmities it has survived, the risks it has run, the mentality that must come with a knowledge that the chicks it tends for so long as eggs and then as leathery, lizardy bodies can have such a low chance of survival if starving times lie ahead.

It may well be that in the winters, when long cold rain-storms drive across the colonies, the genetic imperatives which have encouraged the birds to have wettable feathers, without which they could not catch fish at all, are also the reasons they are dying en masse, even as adults. The shags and the cormorants can have boom years, in which many chicks survive, but just as often they suffer major crashes. It is likely then that many of them die cold and desperate in the rain-soaked plumage which will not dry, Dickensian paupers, shivering to death in winter storms.

Shags, of which there are about 250,000, two-thirds of them in Europe or on its margins, from Turkey to Iceland and Morocco to Ukraine, have stayed on the coast and are probably in slow decline, as ever more people come to use the shores and bring their pollutions with them. Great cormorants, which are bigger and stronger, have done better, with a much wider spread from Greenland to New Zealand, often wintering inland on lakes and marshes as much as on the coast. There are now at least 2 million of them and they may still be increasing. Fishermen and fish-farmers tend to hate them and they are shot, poisoned and drowned to save the fish that human beings have raised and gathered for their own delight. In that way these cousin birds, for so long seen as the embodiment of evil, now occupy a more ambivalent niche with us, emblems of a man-shaped world, both beneficiaries and victims of man's urge to transform his environment. We are making them and unmaking them as we speak. They have no deep ocean resource into which they can withdraw and hide, and it may be that the fate of the shag and the cormorant this century is bound up with ours, with our decisions and our desires, what we want for them and what we want for ourselves.

7

Shearwater

———————————

We had been sailing all night, coming north up the Irish coast from the Aran Islands, four hours on the wheel, four hours off. My watch had started when it was still dark, a cup of tea in the hand, the big westerlies driving in from the Atlantic. As the sun rose, its first touches brightened and reddened the rim and then the body of each wave, revealing the companions that had joined me: Manx shearwaters, not crowds of them but about a dozen, low over the sea surface, fast, lifting and dipping to the swells, flicking the gloss of their black back and wings towards me, then turning away, white, the underbelly, catching the dawn-red reflection from the swells below them. Dark and flushed, dark and flushed: I have never seen anything so beautiful.

They flew a foot or two above the water, as if attached and moulded to it. All morning, our big gaff-rigged cutter plunged and surged for the north and all morning the shearwaters swept and careened beside me as the image of everything I could long for: effortless, liquid, consummate, fitted to their world.

I watched for hours: the way they touch the sea with their wingtips as they turn is the source of their name. Those primaries dipped into the water, first one wing then the other, leave

The master voyagers

a fine trailed wake behind them; no need for a safety margin, but as certain at speed as a motorcyclist taking the corner, his knee grazing and kissing the tarmac. It is an ancient name, which means simply 'cut-water' or 'slice-water'. Occasionally the birds flapped their wings, rose and moved away, but it was that shearing movement to which the eye always returned, the sweep of the body, the calligraphy of beauty, each move as unconsidered as the one before, the birds playing the instrument of themselves.

In the *Odyssey*, Hermes, the boundary-crosser, the musician, the merchant, the trickster and the traveller, the god surely of seabirds, becomes a seabird himself. Zeus sends him to save Odysseus from the arms of Calypso, who is in danger of suffocating him with her relentless and insistent love-making. Odysseus needs his freedom and Hermes is the answer. Homer's word for what the god becomes is *laros*, now used for gulls, but Hermes is no gull since he dives beneath the waves as a gull never would, and the best guess is that he is a shearwater, dipping his dark primaries into the outermost skin of each wave. The messenger god comes 'swooping down from Pieria [the home of the Muses], down from the high clear air, plunging to the sea and skimming the swells, like a *laros* who hunts along the deep and deadly ways of the barren salt sea'.

It is an image that has stayed with the shearwater, which in Pope's translation 'dips his pinions in the deep' and of which some species have always been inhabitants and breeders in the Mediterranean. Journeying is at their core. Homer's Hermes became the model for Virgil's Mercury in the *Aeneid* and, much later, as another bird Milton could use to portray Satan in *Paradise Lost*. Not this time the lugubrious cormorantine tempter of Adam and Eve, but the cosmic wind-chaser, dropping from the heights of heaven in the grandest rendering of seabird flight ever heard in English poetry:

> Mean while the adversary of God and man,
> Satan with thoughts inflamed of highest design,
> Puts on swift wings, and towards the gates of hell
> Explores his solitary flight; some times
> He scours the right hand coast, some times the left,
> Now shaves with level wing the deep, then soars,
> Up to the fiery concave towering high.

Almost every image Milton had of the universe came from the picture of men and creatures out on the expanses of the ocean. The shearwater was born to occupy that vastness. It has distance in its genes and the kernel of its life is not on the breeding islands or in their dark and musty burrows but out in the extent of space, on the breadths of the ocean, vividly alive in that huge and narrow slice of air, a yard or two high, 20,000 miles wide.

Milton's cosmic wind-chaser, the great shearwater

Shearwaters are relatives of the fulmars, the albatrosses and the petrels, all members of the super-family called the tubenoses – they have a pair of exterior nostrils moulded to the upper side of their bills – or the Procellariiformes, an unpronounceable nineteenth-century German term with a beautiful meaning: the Order of the Storm Birds. The Procellariiformes are the emperors of the wind, the greatest of all travellers, the birds which have evolved most entirely to make use of the ocean. Thinly scattered prey and the need to lay an egg in a fixed place require speed of passage and an ability to track that prey from a distance. Many of these birds make enormous annual migrations from one hemisphere to the other. The sooty shearwaters cover 40,000 miles from Australia to Alaska and back every year, their Atlantic cousins scarcely less. The burning question is how.

Investigations into the migratory skills and habits of the tubenoses began with the Manx shearwaters from Skokholm off the Pembrokeshire coast in 1936. David Lack, a supreme birder from boyhood and later to be the great theorist of seabird life, the first to apply Darwinian principles to the question of bird behaviour and life-pattern, had been a brilliant student at Cambridge but was now a schoolmaster at Dartington Hall in Devon. He was writing *The Life of the Robin*, the first natural history book to reveal the ferocious and often murderous rivalry between garden birds. In the summer of 1936, Lack took some of his pupils, including the children of Bertrand Russell and of the architect Clough Williams-Ellis, to visit Ronald Lockley, the pioneering naturalist who had set himself up on Skokholm to study the birds. A decade before, Lockley had taken a lease on the island, 240 bare acres with a ramshackle medieval house tucked away from the wind under a knoll. Six pairs of shearwaters nested in burrows within a few yards of his kitchen. 'I had only to slip out of the door each evening, walk a few paces, and with the aid of a torch and notebook, examine and note the movements of the individuals in this colony.' Lockley, a dreamer but also a pioneer, modestly described himself as 'an amateur field naturalist' but was a person of great ingenuity and in love with the birds. Baggy tweed jacket, ill-fitting trousers, elasticated belt with a snake buckle around his waist, a shock of hair, a man who believed in the individuality of birds, who had made a documentary with Julian Huxley and Alexander Korda about gannets that would win an Oscar the following year, and who had taken a razorbill to Buckingham Palace as a pet for the Princesses Elizabeth and Margaret and had been bitten by it when first revealing it to the girls: here was something of a visionary. He may have been the lessee of Skokholm, he wrote, but the birds were 'ever the true, the constant freeholders'.

Every evening in summer the island was full of the shearwaters' cries, 'a bedlam of weird screaming ... There seemed at first no intelligence in that wild howling – it was the crying of insane spirits wandering without aim or restraint over the rough rocks and the bare pastures.' A friend of his thought the shearwater's cry was 'like the crow of a throaty rooster whose head is chopped off before the last long note has fairly begun'.

The shearwater was virtually unstudied. All the books that Lockley could find concluded their entries for the shearwaters and the other Procellariidae with the words: 'Habits unknown.' There were 10,000 occupied shearwater burrows on Skokholm and from 1933 onwards Lockley started to ring the birds. 'I want facts about individuals not general information,' he wrote, and rings allowed him to study individual birds. They were expensive – three farthings each – and the early rings wore away quickly and had to be replaced at least every other year. Nonetheless, he and his friends and helpers ringed 3,000 or 4,000 shearwaters on Skokholm every year.

Only 3 per cent of rings were ever seen again, but they soon started to turn up on the coasts of Devon, Cornwall, south-western France and northern Spain. The birds had been shot by fishermen, picked up dead on beaches or caught in nets. One ring was found in the belly of a bottom-dwelling angler fish in Brittany. Lockley decided that the angler fish must have eaten another fish which had bitten off the shearwater's leg, as he had seen plenty of Skokholm shearwaters with toes or even legs missing, which he judged the work of fish nibbling from below.

Lack and the children from Dartington set up their bell tents on the island, careful not to put them over any shearwater burrows, and started to explore. Every day was spent clambering among the bluebells and the sea pinks on the cliffs, fishing and paddling in the rock pools. In the evenings, Lockley entranced them over the fire in his house. How does the young cuckoo find its way to Africa alone, late in the summer, long after its parents have flown south? How does the swallow return to its nest? How does the great shearwater find its way back to Tristan da Cunha, that needle in the haystack of the South Atlantic, after six months of wandering in the northern winters, about Greenland and the waters of the Gulf Stream? What is the secret of homing?

Some homing experiments had already been done with pigeons and terns, but Lack suggested that the children should take some Skokholm birds back to Devon to see what would happen when they were released. Lockley collected the shear-waters – 'phlegmatic and uncomprehending of danger' – from the burrows outside his door. They carried rings numbered RV 7569, RS 2255 and AE 699, 'my most cherished and the oldest bird in the knoll', also known as Caroline. By 1936 Caroline was on the third or possibly fourth husband Lockley had known. He was christened Carol III, the previous two, or three, having been Carol, ?Carol and Carol II. (Lockley was never certain that ?Carol was not actually Carol whose ring had fallen off over the winter.) Carol III was left guarding the egg Caroline had laid.

With those three shearwaters, three storm petrels and six puffins, Lack and the children left for Dartington. He and Lockley felt sure that the shearwater egg was likely to be fine under the remaining parent, as shearwaters have a habit of heading out to sea for up to twelve days, only then returning to exchange places with their by-now-starving spouse.

Two of the three shearwaters and two of the storm petrels unaccountably died on the train back to Devon. A day or two later, Lack released the puffins, the remaining storm petrel and the shearwater at Start Point, the great headland on the Channel coast. They flew from his hands at 2 in the afternoon. That evening, at a quarter to midnight, Lockley checked the nests on Skokholm from which he had taken the birds. All the puffins, the storm petrels and two of the shearwaters were still missing but to his amazement Caroline was back on her nest, the egg warm beneath her. Carol III had spent the whole time out at sea, indifferent to his duties. She had taken less than ten hours to return from Devon. When Lack had let her go, she had headed south, out in the English Channel and so her return to Pembrokeshire was probably not directly overland but the long way round, via Land's End, a distance of 225 miles, at an average of 23 miles an hour.

No one had any idea how the bird had done it. It may be that Caroline already knew the geography of Devon and Cornwall and simply found her way home along familiar routes. But 'success with Caroline was provocative'. Lockley and Lack were excited by the results and they embarked on further experiments.

Over that summer and the next, shearwaters were whisked away from Skokholm to an eclectic variety of spots: inland in Worcestershire, Birmingham, Lancashire, County Limerick in Ireland (from a train), on the Isle of May, Stockholm, the Faeroes (widows of the birds that had died en route to Dartington), at sea off Portugal, north of Shetland and off the north-west coast of Africa. Many birds failed to return at all. Some only reappeared the following year. Two birds were sent to America and released near Boston, but both were in weak condition and they did not survive. But one or two repeated the dazzling success of the Devon bird. Two Skokholm shearwaters released at Frensham Ponds in Surrey at dusk on 8 June 1937, some 200 miles from the island, were back in their burrows the following evening. Such a quick return, from a place they had not visited before, can only have meant that they had followed a direct course overland to the island.

As Michael Brooke, the world expert on shearwaters, says, the Frensham birds had clearly 'won their homing spurs'. After a month's rest, Lockley sent the two birds to Venice by plane. Two days after leaving the island, in the early evening of 9 July, in one of the great moments of twentieth-century seabird-ology, the shearwaters were released over the Venetian lagoon, a quarter of a mile south of the Giudecca. Alan Napier, His Majesty's Consul at Venice, there to oversee the opera-tion, reported that the sky was clear, the sun setting, the water of the lagoon dead calm and no more than 'a suspicion of a breeze' coming in from the east. Both birds were dipped in the lagoon water: one headed south, out into the Adriatic, the other west towards the Alps in the rough direction of Skokholm.

Lockley's own photograph of 'the wing that flew from Venice to Skokholm'

One of these birds disappeared. The other was a revelation. On the evening of 23 July, two weeks after she had left the Giudecca, with a moon shining over the island, Lockley found her in her Skokholm burrow. He 'handled and stroked her in the moonlight with something like awe. She was plump and glossy. She had done herself well, in spite of the long voyage.'

Her egg had hatched while she was away and the father had tended and fed the chick so that it too was now 'solid with food and fat'. The entire family was in perfect condition, as if no world-changing event had come anywhere near their lives.

Lockley could only guess at the route she had taken: either a thousand miles north-west over the Alps and then the Vosges, straight over Paris to the Channel and on to Wales; or, perhaps, given the shearwater's intimacy and attachment to the sea, down the Adriatic, around the heel and toe of Italy, across the Mediterranean to the Strait of Gibraltar, around Portugal, turning at Cape St Vincent and then over the Atlantic to Pembrokeshire, a journey of about 3,700 miles. Lockley thought she must have come straight over the European land-mass, 'unable to resist the mysterious "pull" of home', perhaps over the Apennines to Genoa, then across France to the Bay of Biscay, feeding there on sardines, the source of her gloss and well-being, and then home, heading not just for home waters, or the familiar stretch of country, but for that particular Skokholm burrow, for the mate and the chick waiting there. At the end of the following March, the second Venetian bird reappeared on Skokholm, also apparently no worse for its adventures.

Lockley and his shearwaters became famous. In June 1952 Rosario Mazzeo, the clarinettist and personnel manager of the Boston Symphony Orchestra, was one of the latest pilgrims to Skokholm. He was flying back to America in two days' time, and Lockley seized the chance. Previous attempts to take shearwaters to America by ship had failed because the birds had become exhausted and either died on the long sea journey or were too weak on arrival. Two shearwaters were taken from their burrows by Geoffrey Matthews, an ex-RAF bomber navigator, now researching bird migration at Cambridge. Matthews ringed them, put them in a 'fibreboard box, 16×10×10 inches with a diagonal partition, one bird being placed on either side, bedded on dry bracken as in the nest'. Mazzeo took them that night on the sleeper to London.

> The birds caused no little wonder and merriment to the people in the adjoining rooms, who could not understand the origin of the mewing and cackling sounds which came from my room in the late evening. The next day the birds remained in the carton, each in its own compartment, and in the evening I enplaned for America with the birds under my seat. Only one survived the journey.

Safely through customs at Logan airport in Boston, a TWA truck took the clarinettist and his one remaining shearwater, ringed AX6587, to the easternmost point of the perimeter on the edge of Boston Harbor. The TWA man helped Mazzeo get the shearwater from its box and no sooner was it out than AX6587 flew across the water, heading east. It was 8.15 a.m. Eastern Standard Time, 3 June 1952.

Geoffrey Matthews, later to be the man who saved hundreds of millions of acres of the world's wetlands and their birds by setting up the Ramsar Convention, was waiting on Skokholm for news. Every night, he checked the burrows. At 1.30 in the morning of 16 June, as he later wrote to Mazzeo, he was:

> completely flabbergasted to find AX6587 in its own burrow. I read the ring several times, and then put the bird back and blocked the entrance. I wanted to make sure ... As we had not then had your letter, I was convinced that you must have run into trouble with our customs and released the bird at London. The boat came over that morning with your letter – there was no gainsaying the result then! A pretty touch, the bird beating the mail!

AX6587 had travelled at least 3,050 miles in twelve days, averaging 250 miles per day without a single landmark to guide it until it had, presumably, come across the southern coast of Ireland. It was by far the longest homing flight yet recorded and the first time any bird had been tracked across an ocean.

Matthews, with a team of assistants, then embarked on more refined tests. At various points around England, including Cambridge, London and Slimbridge, where he was already beginning to work with Peter Scott on creating a haven for wetland birds, kidnapped Skokholm shearwaters were set free. The best results came from the top of the tower on the University Library in Cambridge. 'The liberators released the birds one at a time, tossing each one up into the air and following it with binoculars until out of sight before releasing the next in a different direction.' Carefully compiling the results from 338 shearwaters transported and released in this way, Matthews established, for the first time, what he called 'evidence of true, bi-co-ordinate navigation'.

At the moment they left the tower and when the sun was visible, the shearwaters orientated themselves towards Skokholm. Some of those Cambridge birds, released in the evening, were back in their burrows 175 miles away by soon after midnight the same night. Two birds, which got back to Skokholm from Cambridge in one night were released again from Haydon Bridge in Northumberland the following year and also managed to return the same night. 'We have here, then, extremely strong, concrete evidence that the shearwaters were at least roughly orientated towards home within about 3 minutes of release, arguing that they must possess some form of navigational mechanism.'

When birds were released under a heavily clouded sky, their idea of where to go broke down completely. Most of them headed away from Skokholm, with the birds hesitating and circling near the tower, uncertain of their direction. Two-thirds of the birds released under a cloudy sky went missing permanently, lost somewhere in the fields and woods of southern England. Birds that had been released at night, even within a few miles of Skokholm, never got home before morning. As soon as the sunlight returned, those night-released birds could head back to their island and were in the burrow by the following evening.

Although it was not clear why there should be this difference between a night sky and a cloudy sky, it was at least obvious that the sun was the key. Matthews and others were conducting parallel experiments with pigeons and lesser black-backed gulls. All three species showed the same responsiveness to the presence or absence of the sun. The sun itself was their compass. It is far enough away for its apparent position not to change as the bird flies across country. Of course, the sun moves across the sky and so the bird has to allow for the time of day. Amazingly, birds do appear capable of making that adjustment. Their internal clock allows them to know the time, relate the position of the sun to that time, calculate where the sun would be at midday and establish south in their minds. Matthews's experiments with pigeons also showed that birds can measure the height of the sun at midday. The higher the noon sun in the northern hemisphere, the further south you must be and so measuring the height of the sun above the horizon allows the bird to calculate its latitude. This was Matthews's 'true, bi-co-ordinate navigation': longitude from the sun position plus a sense of time, latitude from sun altitude.

The use of the sun compass was one thing, but a compass is little good without a map. The bird might be able, almost instantaneously, to make these extraordinary calculations to establish its position and orientation, but how did it know where to go? How did the Skokholm shearwaters released from the University Library in Cambridge know where Skokholm was?

As the shearwaters never returned to their burrows in the daylight, they clearly had other means of navigating. The sun may have brought them to the area of Skokholm but something else must bring them to the burrow. Their night vision is poor, and they often blunder into lighthouses, each other and seabird scientists. But in the dark other senses come into play. The birds usually gather at dusk in crowded floating rafts offshore, waiting for the night to cover their return to land. Their legs are set so far back on their bodies – to help them dive – that they are stumbling incompetents on land, exceptionally vulnerable to predators. Darkness, like the burrow itself, is a protection against the clumsiness which a devotion to life at sea has burdened them with. And as they come ashore, they begin their extraordinarily loud and raucous calling, the male shearwaters distinct from the female, and in fact all shear-waters distinct from all others. It brought back memories to Geoffrey Matthews, who had been with Bomber Command in the Far East: 'Indeed, despite the recognizable individuality of the flight calls, the resultant cacophony may be akin to that produced by traffic in eastern cities, expressive of a desire to avoid collision, of nervous tension and general excitement.'

Maybe it was the calling of the shearwaters that drew them in, at least over the last few yards to the particular burrow, and if so, that would give a function to the raft: the birds had flown in during the very last of the daylight and having gathered there could make their way on the final stretch of the journey to the mate, egg and chick using the sounds of the seabird's cry. But that would hardly have an effect beyond a few hundred yards. The heart of the question remained unanswered for fifty years: what was the language of the shearwaters' map? What allowed them to fly directly from Cambridge over unknown territory, from Venice in a vast circuitous route which would have involved flying for hundreds of miles in the wrong direction, or from the hands of the TWA man at Logan Airport for over 3,000 miles straight back to Skokholm? How did shearwaters understand the nature of their world?

Before the great revolution in miniature electronics that began with albatross experiments in the early 1990s, hints were gathering that the shearwaters were not merely local birds. Lockley himself oversaw the ringing of over 40,000 birds on Skokholm and greedily drank in any information the ringing recoveries could give him. Most of them were from British coasts and the shores of the Bay of Biscay, with one or two stretching down to the Canaries and the west coast of North Africa. Then, in October 1952, even as Geoffrey Matthews was conducting his experiments in England, something revolutionary turned up: one of Lockley's shearwaters ringed on Skokholm was recovered from the southern shores of the River Plate, south-east of Buenos Aires in Argentina. It had been ringed sixteen days before and was thought by its recoverer to have been dead for three days when found: the bird had flown 6,000 miles in less than two weeks. The most astonishing part of this discovery was that the egg in which its life had begun had been laid no more than four months previously. It had been out of the egg for about three months by the time it arrived in South America. The ocean traveller was this year's fledgling.

There had been several earlier suggestions that shearwaters migrated around the oceans on circular routes taking advantage of prevailing winds, but only in the late 1980s was the pattern of their lives first properly guessed at. Michael Brooke, Curator of Ornithology at Cambridge University, collated and examined all 3,600 rings by then recovered from dead shearwaters across the Atlantic. It has been said that trying to puzzle out the habits of a seabird from ringing recoveries is like trying to understand the social life of a city from a map of its murder victims, but Brooke's analysis suggested an intriguing pattern: the fledglings headed south from Europe, along the French and Spanish coasts, to Madeira and the Canaries until they picked up the north-east trades which blew them to Brazil. There, Brooke reckoned, they spent the winter and maybe much of the following year before heading north to the seas of the Grand Banks off Newfoundland, and from there with the westerlies across the Atlantic to the colonies in Europe. This track, looped across two hemispheres of the Atlantic, then became the life-pattern of the bird, exchanging northern winter for southern summer, and southern winter for northern summer in a vast yearly odyssey.

It was an impressive piece of detective work, for which there could be no confirmation until instruments had been miniaturized to the extent that Manx shearwaters could carry them without debilitating effects. That had to wait until 2007, when a team led by Tim Guilford at Oxford recovered the twelve light-logging geolocators, each weighing less than 1/100th of an ounce, from six pairs of Pembrokeshire shear-waters to which they had fitted them the year before. These geolocators measure day length, which gives a latitudinal fix, and have an internal clock, recording the time of midday, which gives longitude. They are not particularly accurate – errors of 100 miles or more show flights overland which are highly unlikely – but the picture was clear enough.

Brooke had been largely right: the autumn birds swung fast down the west coast of Europe and Africa, took to the trade winds, crossed the Equator and made for Brazil before heading south to overwinter among the rich fishing grounds off the coast of Patagonia (from which there had been no ringing returns). The journey north headed for the Equator and then curved west through the Caribbean, some birds reaching the eastern seaboard of the United States and all then heading back east across the North Atlantic. Everything Homer, Virgil and Milton had imagined for the shearwaters was recorded by the geolocators: vastness, command, oceanic extent and speed. The fastest, a male from Skomer, had flown south in the autumn of 2006 covering more than 4,800 miles in six and a half days.

These journeys were moulded to the wind. The routes taken by the birds reflected almost exactly many of the voyages made by the first European navigators heading south into the Atlantic, discovering Madeira, the Canaries, the Azores and then travelling on the trade winds to the continental mass of the Americas before coming home in the westerlies.

The wind is not some capricious element of life at sea but its iron discipline. The vast gyres circulating around each ocean basin, clockwise in the northern hemisphere, anti-clockwise in the south, dictate to their inhabitants the entire pattern of life. But the form of knowledge raised by that idea is almost impossibly opaque. How do the shearwaters know the shape of the gyre? How can they know that heading south along the shore of western Europe will bring them to the trade winds that will take them to the fishing grounds of South America? How can they know that heading north across the Caribbean will bring them in the end to the westerlies off Newfoundland that will blow them back to Pembrokeshire?

The precision is astonishing. Ángel M. Felicísimo, from the Universidad de Extremadura, in Cáceres in Spain, is a specialist in remote sensing technology and the influence of the wind on all forms of life. He created two parallel maps of the South Atlantic for a three-week period in November: one from NASA wind data of what he called 'wind highways' or 'minimum cost corridors' – those routes in which the wind would give most help to any migrating birds; the other from geo-locators attached to the legs of Cory's and Balearic shear-waters from the Canaries, the Azores and Pantaleu, a little uninhabited islet off the coast of Majorca in the Mediterranean. These birds perform miracles of annual migration at least equivalent to anything in the Welsh and Irish shearwaters, but no one had realized until Felicísimo's work quite how much their flyways were bound to the workings of the wind.

Shearwaters from the Mediterranean and the Canaries flew south on a narrow corridor just west of the African coast. Those from the Azores, equally concentrated, took a route a little further out into the Atlantic. Once south of the Equator, they all joined the great wind highway, circling around the enormous ocean gyre, some remaining off the coast of South America, others turning east for the rich fishing grounds off Southern Africa. They were covering between 10,000 and 20,000 miles each year before returning to their breeding colonies in the north.

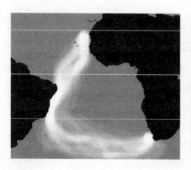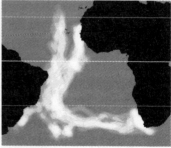

The routes of migrating shearwaters (right) use the Atlantic 'wind highways' (left)

The shearwaters are subtler than you might imagine and there is an interplay here between their knowledge and under-standing of the winds, and an adaptability, an elasticity of response which means they can change their behaviour from one year to the next. Not only do the shearwaters know how the wind works; they are able to climb on and off the circulatory gyres like traffic on a freeway, pulling into rich fishing grounds or continuing across the ocean, apparently at will.

In the next generation of these tracking studies, two Portuguese seabird biologists, Maria Dias and José Granadeiro from Lisbon, have started to reveal flexibility in the migration strategies of individual birds. They are not shut in the prison of the winds but can choose what to do from one year to the next. They have tracked seventy-two different migrations of Cory's shearwaters, and it appears that each bird can select as it wishes, shifting from the South to North Atlantic, from the western to eastern South Atlantic, and even from the Atlantic to the Indian Ocean, spending longer or shorter times in particular places, leaving and arriving earlier or later. These are not machines and not robotic, but decision-making individuals.

Dias and Granadeiro tracked the birds from Selvagem Grande, a tiny and lonely island halfway between Madeira and the Canaries, the great headquarters of Cory's shearwaters, vast numbers of them nesting on what Ronald Lockley called 'a few scorched acres of volcanic rock'. He had dreamed of living here too and had arrived one year to ring a few hundred of these shearwaters and to see if any Manx shearwaters bred this far south. Lockley and his party arrived in heavy fog, the island invisible to captain or pilot. Only the 'splendid hullabaloo' of hundreds of thousands of the birds had acted as a fog signal and guided the ship into the shore.

Cory's shearwaters do not breed before they are nine years old and their long adolescence is spent exploring the oceans. Dias and Granadeiro followed one immature bird, a four- to five-year-old over two successive years, in the course of which it visited every one of the species's overwintering sites: the rich waters off Patagonia, the central South Atlantic, the rich cold currents off the two coasts of Southern Africa, the North-west Atlantic and the Canary Current off the coast of North-west Africa. This 67,000-mile journey looks like a research trip, a gathering of mental resources for the future.

The mature birds, all of which would have been through this oceanic education, then picked and chose between options in their later life, sometimes staying in the north, sometimes in the south, sometimes heading to the Mid-Atlantic Ridge, at others choosing Patagonia. Some alternated between Patagonia and the Benguela Current off Southern Africa, others went for the central South Atlantic or the coast of South America. Birds could winter almost 4,500 miles apart in different years, in different hemispheres, equally at home.

It is difficult to credit other creatures with a decision-making capability on such an enormous scale. Few human beings can live like this, but a magnificent and impartial power of choice is embedded in the shearwaters. These were not random selections, a chaotic veering from one part of the ocean to another. Individual birds chose the same areas more often than would be expected by chance. The Cory's shear-water is both constant and flexible, subtle and inventive, reso-lute and resourceful, an Odysseus-like model for how to be: if they are pushed they bend, but they bend back. If so minded, they stick to what they know. If they want to change their winter destination, they will often stop off at alternatives en route.

The sun may have provided a compass for these dazzling journeys. The knowledge and memory of local coastal condi-tions may have provided at least part of a map. Understanding the nature of the winds and acting according to quite a simple principle such as 'Keep the wind behind you, or at least abaft of the beam' will have delivered shearwaters to many of the destin-ations on their great hemispheric figure-of-eight loops. But none of this is good enough to provide a comprehensive map.

For a long time, it was guessed that these ocean birds could somehow understand the magnetic field of the earth and use that as a form of map. Experienced homing pigeons which have had magnets attached to their heads have been unable to find their way home until the magnets were removed. But equivalent experiments with albatrosses and shearwaters have had no effect at all. It seems that magnetism plays no part in their giant voyages.

It may be inappropriate to look for a 'map'. Evolution, as the French biologist François Jacob famously said, 'proceeds like a tinkerer who, during millions of years, has slowly modified his products, retouching, cutting, lengthening, using all opportunities to transform and create'. The method by which the birds know the world probably does not look like the clean-cut model of compass-and-map. Different senses and different combinations of senses may be in constant conversation with each other, in a shifting menu of variable contributions. And one element among these is almost certainly scent and the sense of smell.

All the Procellariiformes live for a long time, four to six decades, faithful to site and mate; all of them are strongly scented, with a smell which permeates body, plumage, egg and nest; and all of them are equipped with the exaggerated nostrils in the tubenoses above their bills which gives the family its name. They are not unique among seabirds in being smelly: as Julie Hagelin, the Alaskan seabird scientist now based at Swarthmore, has explained, the delicious tangerine scent of the crested auklet, the acrid odour of the variable pitohuis, whose feathers are soaked in toxic, self-protective oils, the sweet and dusky fragrance of the kakapo, the flightless parrot from New Zealand, and the foul stench of the hoatzin, also known as the stink bird of the Amazon, can all be added to the musky plumage of the storm petrels, the shearwaters and other tubenoses as smell-gestures just as flamboyant as the plumage of any lyre-bird or cockatoo.

Smell is of exceptional importance to the Procellariiformes, and the olfactory bulb, that part of the brain which processes scent, is particularly large in them. Just as eagles and other raptors have densely packed receptors in their eyes for intense and acute vision, the shearwaters, fulmars, petrels and albatrosses have up to six times as many smelling cells in that olfactory bulb as, for example, laboratory mice.

Gabrielle Nevitt, now a professor at the University of California at Davis, has been pursuing the significance of scent in seabirds for more than quarter of a century. Her first Antarctic research cruise in 1991, as she told the science journalist Nancy Averett, involved hundreds of boxes of super-absorbent tampons. She soaked them in strong, fishy-smelling juices, attached them to kites which she then flew off the stern of the research vessel hoping to attract the birds. 'On the day Nevitt ran her experiment,' Averett has written,

> dozens of them swooped in so close that [Nevitt] feared they would tangle in the line and drown. So she grounded the kites and improvised, releasing vegetable oil into the water, some of it laced with the fishy compounds. Albatrosses and petrels flocked to the stinky slicks. She was ecstatic. But she still had no idea how they used olfactory cues to home in on their ephemeral quarry. 'I was really passionate about figuring this out, so I wasn't giving up,' says Nevitt. 'I knew I'd be back again soon on another cruise.'

A year later, she met Tim Bates, a chemist at NOAA, the National Oceanic and Atmospheric Administration. He was investigating a gas called dimethyl sulphide, or DMS, which is emitted by phytoplankton, the microscopic plants that live on the surface of the ocean, particularly when they are damaged. Bates was interested in the gas because it can provide tiny aerosols around which clouds gather. More DMS, Bates thought, might mean more clouds, which would prevent more of the sun's rays reaching the earth's surface, which would slow down the process of global warming.

Nevitt was talking to Bates as he was calibrating his equipment. It was a chance coming together of two sciences. From the gas wafting around the flasks and test tubes, Nevitt picked up a smell that reminded her of newly opened oysters. 'She felt a tingle of excitement,' Averett wrote.

> She knew that the gas is released when krill – a major food source for seabirds – devour phytoplankton. 'I'd read about DMS,' she says. 'But it never occurred to me it might have an odor.'
>
> It all clicked into place. The birds pick up the DMS trail and follow it to schools of krill. When Bates showed her a map of DMS plumes, Nevitt saw that they were more concentrated in areas with geographic formations near the ocean's surface. 'I could see peaks and valleys of DMS over shelf breaks, seamounts, and other underwater features, and I realized the ocean's surface wasn't featureless to the birds,' she says. 'They have their own map, an odor landscape, in the air above the water.' It was, says Nevitt, the kind of 'aha' moment scientists live for.

The phytoplankton was growing most densely where the ocean was churned up by roughness in the sea floor. Where the phytoplankton was thickest, the krill was gathering to eat it. Where the krill ate it, plumes of DMS were released into the atmosphere. The seabirds could smell the DMS, read it as a signal for the presence of krill and of fish feeding on that krill. DMS was a flag for food. When Nevitt went to sea and tested the theory that birds would be drawn to DMS released into the atmosphere, it worked: the tubenoses made for any hint of the invisible and transparent gas she released into the air over the Southern Ocean.

Subtleties emerged: the smaller and more vulnerable Procellariiformes tended to find their way to the source of DMS 'using a zigzag, upwind search behavior'. But they didn't appear when Nevitt released the smells associated with crushed krill or shrimp, even though krill is an important part of their diet. Those smells are full of pyrazines, the chemicals at the root of the peppery taste in a very green sauvignon blanc or in cabernet franc. The pyrazines drew in the bigger and more aggressive tubenoses, precisely the birds which would prey on the fish that also ate the krill.

With every passing year, more evidence appears that smell shapes the conceptual world of the shearwaters and their cousins. The burrow in which most of them are born – and the ancestors of fulmars and albatrosses were probably also burrow-nesters – is an odour chamber. The returning parents, bringing food to the chick, smell of DMS themselves and so the association between DMS and home, food, safety and well-being is instilled in the bird in the first weeks of life. A shearwater chick will be seventy days old or more before it emerges into the light for its first journey and so those first seventy days are spent enveloped in a DMS fog. Nothing is more educated than its sense of smell, and in later life, as a whole sequence of experiments in the last few years has shown, smell is implicated in both parenting and pair-bonding: the nibbling and allo-preening on birds in the nest is above all an odour experience. Birds can recognize their own smell, the smell of relatives and that of unrelated birds like them. Their hearts beat faster when they are exposed to smells they know.

Gabrielle Nevitt and her co-workers have tested the sensitivity of petrel nostrils and it has turned out that the birds can detect the presence of DMS in air at a concentration of less than one part in 10 billion. That is the sort of density at which a plume of DMS would be naturally present above the ocean. If storm petrel chicks have their smelling organs temporarily disabled with a chemical wash, they are unable to return to their burrow even from 3 feet away. Smell may well be the governing sense of seabirds. It is the most ancient of senses, evolving with the first brains before the eye had appeared, one of the key instruments for predators and prey in the first arms race that erupted in 'the Cambrian bloodbath' of 540 million years ago. So ancient is the sense of smell that the female sex pheromone in Asian elephants is chemically identical to the same pheromone in many moths.

Because smell plays a small and marginal part in our intensely visual lives, we humans tend to underestimate its importance. But smell runs deep. Lucia Jacobs, professor of psychology at Berkeley and one of the world's leading researchers into the role of smell in navigation, has proposed the idea that the sense of smell – 'the olfactory system' – originally developed not to distinguish between objects that smelled different but so that animals could navigate around an unsettled world. She quotes the neurobiologist Cornelia Bargmann:

> The visual system and auditory system are stable because
> light and sound are immutable physical entities. By contrast,
> the olfactory system, like the immune system, tracks a
> moving world of cues generated by other organisms, and
> must constantly generate, test, and discard receptor genes
> and coding strategies over evolutionary time.

The sense of smell has a genius for locking on to subtle targets. Jacobs has shown experimentally that people can find their way back to a particular spot by smell alone. She first put sponges infused with two of three smelly oils – sweet birch oil, anise oil and clove bud oil – in two places around a room at her university in Berkeley, California. One at a time, she took forty-five undergraduate students into the room, disorientated them by turning them round and round and then placed them in a target location for a minute. With each student, she did this three times: once blindfold and wearing earplugs and noise-reducing earphones but with their nostrils open; once unable to see, hear or smell anything; and once able to see but with the other senses blocked. Each time, they were then led around the room for a while and taken back to the door they had come in by. Could they now find the target location again?

The best results were when the students could see: they could find their way back to where they had been with no difficulty. When all senses were blocked, the students were able to get no nearer than 12 feet from the target. But when they could smell their way back towards it, even though they could neither see nor hear, they were on average 9 feet 6 inches away. That is good enough for science; it is a statistically significant difference; and this remarkable result, as extraordinary in its way as the discovery of echo-location in bats or the ability of sea turtles to orientate themselves in magnetic fields, showed that humans can navigate to an arbitrary location whose position they have learned using only their noses.

Are the shearwaters doing what the Berkeley students were doing? Have they also been able to embed in their minds a smell map of the ocean, not simply zigzagging upwind to an inviting source of prey but another navigational step on from that: a calculation of position and route from the different intersecting plumes of chemicals blowing across the face of the ocean?

In the summers of 2010 and 2011, Anna Gagliardo from the University of Pisa wanted to ask this question of some Cory's shearwaters on the Azores. She chose the colony that has come to settle on the most recent of Azorean volcanoes, at Capelinhos on the western tip of Faial, which began to boil and burst up out of the wonderful, deep mid-ocean blue of the Atlantic in September 1957.

In each of her two years, Gagliardo took twelve shearwaters from their nests. She wanted birds that were not hungry and were highly motivated to return to the nest and so only those that had come back to the colony during the previous two nights were captured. Their eggs were replaced with plaster eggs and kept in an incubator until either parent returned. If the experimental birds did not come back or their mates abandoned the eggs, Gagliardo gave the eggs to shearwaters whose eggs were already broken or addled (all this a measure of the ethical improvement since the days and practices of Ronald Lockley on Skokholm).

The scientists carefully pushed a curved needle with a rounded tip into each nostril of eight shearwaters. Half a teaspoon of zinc sulphate solution was washed over the olfactory cells inside. This has the effect of temporarily depriving the birds of their sense of smell. The cells regenerate within a few weeks, but for the duration of the experiment these birds for the first time in their lives were unable to smell anything. Eight other shearwaters had a small PVC box glued on to their head feathers. It held a tiny magnet weighing about one-eighth of an ounce. The magnet was free to tumble about inside the box so that the bird would be constantly surrounded by a varying magnetic field. The remaining eight shearwaters, like the other sixteen, were simply fitted with location-loggers.

Gagliardo was going to test whether deprivation of a sense of smell or of an ability to respond to the earth's magnetic field would have any effect on the way they would navigate across the open ocean.

She knew that tubenoses deprived of their sense of smell have difficulty in finding their burrows. There was a danger she would not get the data back from them and so those birds were equipped with PTT transmitters which send their position direct to a satellite, with no need to recover the tag. Five of these eight birds had solar-powered transmitters attached to their backs with a Teflon ribbon harness crossed on the breast. The other three were battery-powered, fixed to the back feathers with water-resistant tape. The shearwaters also wore GPS tags. All this may sound rather a long way from the world of Ronald Lockley and David Lack, but each of Anna Gagliardo's birds, apparently equipped for a heavy-duty Himalayan expedition, carried a payload of just over an ounce.

On the evening of the day they were caught, separate cardboard boxes containing one bird each were embarked on the *Monte Brasil*, a 7,000-ton container ship en route to Lisbon. A day later, 500 miles east of the colony and more than 300 miles from the easternmost island of the Azores, the birds were released, lifting away at twenty-minute intervals from the vast metal bulk of the *Monte Brasil*. The biologists were careful to send them out in turn: a control bird, with no manipulation of its senses; a demagnetized bird; and then one with no sense of smell.

Anna Gagliardo's results have transformed our understanding of seabird navigation. All the control shearwaters flew back to the breeding colony almost without hesitation or deviation. The magnetized birds performed in nearly the same way. But of those temporarily deprived of their sense of smell, only two of the eight were able to find their way back home that year to the dusty volcanic slopes of Capelinhos in Faial.

The control and magnetic birds were clearly directed home-wards, but the non-smelling shearwaters wandered across the ocean for thousands of miles, profoundly lost. Their tracks on the chart of the ocean are almost unbearably poignant, wobbling and quavering, circling in search of a knowledge of where to go, desperate voyages, 4,000 miles or more, before disappearing from view a month or six weeks after they had been released from the taffrail of the *Monte Brasil*. Each looks like the route of a creature that has lost its mind, stumbling up into the outer reaches of the Bay of Biscay, prospecting off the coast of Portugal, quite literally wondering where it was. Their journeys look like the searching of a blinded child. Dr Gagliardo did not measure the levels of stress hormone in these birds, but it is inconceivable that they did not suffer the anxieties of the missing and the lost.

The birds that made their way straight home cannot have been following an odour plume blowing towards them from the colony. These are the horse latitudes, of uncertain winds, and frequent calms, intermediate between the trade winds to the south and the westerlies to the north. Both the control and the magnetized shearwaters, as soon as they left the stern of the ship, turned for home, 500 miles away, in winds that were not blowing from there.

The only viable explanation, miraculous as it might be, is that different parts of the oceans are remembered by the birds on the basis of what they smell like. They have flown this way before, and they remember the sea as we might remember places we have known, a rich and various landscape, but one which happens to be invisible – unsmellable – to us. Perhaps it is a dimethyl sulphide landscape; perhaps other chemicals contribute to it. In Gagliardo's experiment, two of the shear-

waters deprived of their sense of smell came back home after
three weeks wandering across the North-eastern Atlantic. A
third returned three months later, probably after its nostrils
had recovered. Others remained lost.

Shearwaters and the other ocean navigators do not use a
magnetic map: they smell their way around the ocean. But
more than that they can smell how the sea works, where the
fish are, the life of the sea. They can smell their way down the
links and layers of the food web, into the presence or absence
of the plankton on which they and all other sea life depend.
They are not only barometers of the sea but its investigators
and navigators. What may be featureless to us, a waste of
undifferentiated ocean, is for them rich with distinction and
variety, a fissured and wrinkled landscape, dense in patches,
thin in others, a rolling olfactory prairie of the desired and the
desirable, mottled and unreliable, speckled with life, streaky
with pleasures and dangers, marbled and flecked, its riches

Plastic in the guts, mistaken for food

often hidden and always mobile, but filled with places that are pregnant with life and possibility.

Intuitively, and at enormous depths in human history, people have always understood the miraculous capacity of the seabird to thrive in a place which human beings find almost impossibly difficult. With noses that can smell, enlarged olfactory bulbs that can detect the subtlest of differences, the ocean becomes a parkland of meaning, alive with signals telling you where you are and where you need to go, a miracle of otherness, an ocean planet of which until now we have been almost completely unaware.

There is one disturbing footnote to this remarkable sequence of discoveries: small pieces of plastic that have been floating in the ocean for a while also produce a plume of dimethyl sulphide which all of the tubenoses – the shearwaters, the fulmars, the petrels and the albatrosses – now mistake for food. They don't eat the little bits of plastic because they look like prey but because they smell like it. The plastic we have scattered across the ocean has become a sensory trap for the birds, and other sea creatures, that are out there in search of sustenance.

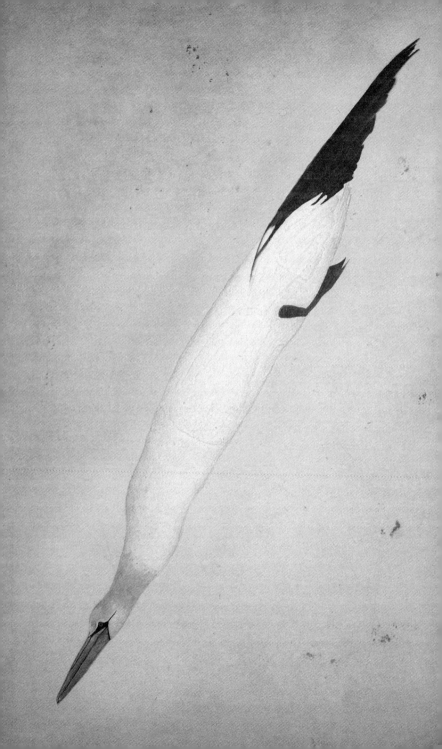

8

Gannet

Halfway through a summer Sunday on the Gower Peninsula in South Wales, the tide was pulling back from the sands around the big silting estuary of the River Loughor. The cast-iron tower of Whiteford Point light, rusting, panelled, armoured, was standing with its feet in the shallows of the sea. People were on holiday and the caravan parks were full.

Mike Buckland, a welder from Heath in North Cardiff, a curly-haired, quietly funny man, who had spent his life making gates and repairing the chassis of broken cars in various garages, staring through the dark of a welder's mask, was down there with his friend Alison Stanley. They often came to the caravan she kept near Llangenneth overlooking the sands.

The two of them were walking on the beach. The tide was way out and the sands lay flat and wet, a sodden mirror continuous with the sea. They saw the gannet sitting in a patch of seaweed, about 100 yards in from the tide, upright, looking about itself, enormous, its head 2 feet above the weed, the white of its feathers a blazing signal of its presence, the rim of pale blue around its eye unlike any other colour south of the Arctic.

The bird could neither walk nor fly and when Mike and Alison came up to it they saw blood beneath its wing and what might have been a bone sticking out through feather and flesh. 'It must have mistaken the sand for the sea,' Mike said. 'The sun must have glinted on the beach. It lies so wet there and the beach is like a sheet of steel when you get the angle right. It looks like a mirage. Or even when the tide is in, you can walk three-quarters of a mile out with the water only up to your waist.'

The bird had come from Grassholm, a gannet citadel 60 miles to the west beyond the Pembrokeshire headlands. Nearly 120,000 gannets spend their summer there, and crowds of them had come in now to the Welsh estuaries, a flight of little more than an hour from the widths of the Celtic Sea.

When the birds arrive at what they think might be a shoal of fish, they make two kinds of dive: the first exploratory, dropping steeply, quick in and quick out, a V-shaped trajectory, leaving vapour trails inside the sea, a detonation of in-carried air. They test whether the hint in the sea, the guessed-at body, is the thing itself. You have to dive to know. Only then, once the picture is clear, do the gannets begin to fish in earnest, U-shaped dives this time, plunging into the water 10 or 12 feet from the impetus, and then swimming with their half-closed wings, pulse-pursuing the prey, which they either swallow whole before rising or bring to the surface to eat.

The Gower beach had cheated this gannet. Possibly, the wet sand was dancing and flaring in his eyes, far below, as if full of life, sprinkled with the sheen of living bodies, made mobile and alluring by the bird's own passage across it. Or the sea in which he saw the fish was no more than a foot or two deep. Gannets can dive in such shallow water, but this one may have mistaken the depth, and so made his dive, from 60, 70 feet, folding, arrowing, transforming himself into a one-point trident, reaching 40 or 50 miles an hour, smack into the concrete of the sand. Or it might have been wounded out at sea.

Mike picked the gannet up, thinking he might move it to the dunes at the head of the beach, where he had seen a dead sheep lying in the marram grass. 'I thought I could put him there so he could tuck in,' he laughs. He had handled many birds before. 'When I have seen animals I have always picked them up, even as a boy. If I found a sparrow I would keep it in a box in the kitchen, feed it worms. The best one was a raven. I found it in a quarry. A fox was chasing it. I only had it an hour before it was sitting on my hand. I came home with it on the bus, a bus full of people, all very happy. And it stayed with me a couple of months before it flew off.'

He had never seen a gannet before. 'But I grabbed on to the beak when I picked it up. It was fine. It did flutter a bit at first but it was soon all settled in after it had done its flapping about. It was in my arms. I think it was enjoying the walk. We all were. Its head was bobbing along looking at the beach ahead. It was content. Well, it probably wasn't content but it looked content.'

A family of four, on the beach with their small black dog, came across the sand towards Mike and Alison and their bird. The dog ran towards them and barked, jumping up trying to get the tail feathers and so Mike moved the bird up a little. He looked down at it, still tucked under his arm like a bagpipe, to see it was not troubled, and as he did so the gannet reached up and destroyed his right eye, cutting away with the razortip of its beak at the eyeball, at Mike's nose and then at his other eye. The pain, Mike said, was 'like somebody was sticking a pin through the back of your eye and trying to prise it out'.

'I put my right hand up to my face and I felt like there was a big hole there where my eye was meant to be. It wasn't hanging out but it was down inside. A bit of a push and it seemed to pop back in. But I didn't feel it was there, until I felt it at the side. The bird was cutting my nose then. I had a gash, a flap of skin, across the top of my nose. My eyelid was ripped off on my left eye. But if I didn't blink in time, I would have lost that eye as well.'

The family with the dog disappeared. 'They did say, "Are you OK?", but I couldn't see anything and I just turned in the other direction. Alison was screaming. I wouldn't let her see my face until I had popped my eye back in. I just walked with the bird. I didn't let it go. She was saying, "Put it down, put it down," but I just walked in a dead straight line up the beach and then I stopped.'

And then?

'Well, after the shock, I took the bird up to the sand dunes, it wasn't far. I couldn't see where I was going but Alison was with me. But he didn't like me then, and he locked on to my hands with his bill and I had to shake him off. That did cut my hands a bit.' The inner edges of a gannet's bill are gently serrated. 'Then we left him there and walked back to the cara- van site. A good mile, forty-five minutes. After the shock had gone, I was joking with Alison. It was a weird situation. I couldn't see anything so I was cracking jokes. I got used to the pain but I wanted a cigarette and I kept asking Alison if I was lighting the right end.'

A week after the incident, Mike and Alison went back to find the bird. It was a hundred yards from where they had left it, lying dead on the sand. Would he pick a gannet up again? 'I would have done it again the same day. I've always liked wild creatures.' He has not been back to work since and, although in his mid-forties, thinks he never will. 'You need two good eyes to do any work like that. I'll probably just potter about. I didn't even know what a gannet was. They asked me in the hospital what kind of bird it was and I said it was a giant seagull. I know a lot about them now. I had a quite a good close up of the bill.'

* * *

We have forgotten the rawness of these animals. We want to imagine that their inner beings are as beautiful as their appearance in the world; or that they might have the same attitude to us as we, in our late, careful, conservationist way, have towards them. The best of us, if we are lucky, will behave like Mike Buckland with his gentle, generous, attentive tolerance. But we fantasize a global love and forget that unaccommodation, the refusal to allow, anti-forgiveness and anti-understanding are the foundations of a seabird's world.

If you look into the pale ice-blue blank of a gannet's eye, you see something beyond our world. They act to their own principles and are not to be understood as decorative or charming elements in ours. They live in a violently difficult environment. Their dictates are both strange to us and weirdly familiar, versions of our own habits translated into a language we will only ever struggle to understand. They are our co-beings, co-dominators, co-predators, and like us they embody genius and obstinacy, elegance and terror, persistence and aggression, and perhaps they entrance us because they present us with a vision of the worst and best of ourselves.

The pitiless eye

There is one bird, though, for which cruelty to its own kind has become embedded in its entire existence. The Nazca booby is a beauty, an inhabitant of the Galápagos and a few islands nearer the South American shore, a relative of the gannet translated to the tropics. For the northern gannet's cool Arctic colours, the Nazca booby has substituted the glamour of the south: flecked orange eyes, a bill the colour of ripe mango, with chocolate-brown tips and broad trailing edges to the white of its wings.

These mesmerically beautiful birds – 'booby' means 'idiot', the verdict of early European sailors who thought them stupid as they had no fear of men – are plunge-divers like their northern cousins, specialist hunters ranging wide across the tropical seas for shoals of sardines and flying fish. But their life is one of comprehensive dysfunction. On the dry desert islands of the Galápagos, boobies nest close together in colonies of several thousand birds. They lay two eggs, from three days to a week apart. If both eggs hatch, the elder chick will always kill its sibling, usually pushing it out of the nest, where it will then die of thirst or starvation. The parent never intervenes, but sits sheltering the murdering chick while the victim is shoved into oblivion. A murderous fight with a brother or sister is how every Nazca booby life begins.

Other birds behave like this, the kittiwakes as well as some of the eagles, ospreys and pelicans, but what happens next in the life of the Nazca booby is unique to it. David Anderson, a professor at Wake Forest, North Carolina, has been studying the birds of Isla Española in the Galápagos since 1985. Every winter and early summer he and his research students have seen the same pattern of behaviour developing. Once the surviving chicks have grown up a little, their parents feel able to neglect them and go fishing, almost always in the afternoon. That is the opportunity the young unemployed boobies, the pre-breeding adolescents, up to ten years old, have been waiting for. Every afternoon they turn into what Anderson has called Non-parental Adult Visitors or Navs for short.

Navs patrol the colony. There are more males than females in the booby population and so most Navs tend to be young unemployed males. When they spot an unguarded chick across the flat, open ground of the colony, they fix their stare and set off determinedly towards it, looking hard at the target all the way. The very young chicks aren't bothered by them because they are still protected by their parents and those near to fledging aren't troubled either because they can defend them-selves. It is the mid-stage children, big enough to be interest-ing, not so big that they reject the advances of the Navs, that receive the attention. Chicks never approached Navs.

The Nav's unwelcome visit usually lasts about ten minutes but can go on for an intolerable hour. The visitor sometimes simply stands by the chick, enjoying being near it, giving it presents of pebbles, sticks and feathers, gently touching and preening it just as a mother would with her own chick. These grooming love moments can develop into something far more aggressive. The Nav starts to bite and jab at the chick's body, usually the neck, the back of the head or the wing, clamping its bill around the neck and violently shaking the chick or pull-ing on its down. Finally, the Nav's visit climaxes in an attempt to have sex with the chick. The chick cowers, tucking in its wings and legs tight against body, its bill deeply underneath its chest, submissively presenting no threat to the attacker, while the Nav stands on its back and paddles away with its feet, look-ing for 'cloacal contact'. Sometimes, in the hottest part of the afternoon, the chick will overheat, crushed close to the stony ground and will raise itself to cool down. That summons more extreme aggression.

Over eight seasons Anderson observed 4,945 Nav events in a small sub-colony of about 450 nests. Half the time the male Navs were aggressive, a third of the time they were loving, and for one in seven cases they tried to have sex with the chick. The females were less sexual but often more violently aggressive. When the females were sexual with chicks they mimicked male rape. Most chicks that lived until they were twenty-one days old had been attacked at least once. By the time they were sixty days old, the Nav visits were declining, and when they were ninety days old the chick could resist the molester.

Nav visits only happen when the chicks' parents are away. If a parent returns to the nest, the Nav immediately runs away, escaping the advancing attack by the enraged mother or father. Almost always the chick is left deeply scarred and, particularly in dry years, when food is short, the carnivorous Galápagos mockingbird perches on the chick, pecking at the wounds, deepening them, drinking the blood and eating gobbets of flesh from inside the chick's body. 'Chicks were increasingly traumatized by the blood feedings as the hole deepened,' Anderson has written, 'and generally died a day or so after the blood feeding began, often wandering aimlessly around the colony and dying far from their nest.' Up to quarter of all Nazca booby chicks die in this way.

Does this syndrome have some kind of selective advantage for the genes the Nazca boobies carry? Is this a kind of genetically determined boot camp for a particularly tough fishing existence? Do the boobies need to maltreat their young to prepare them for life, exposing them to the sort of hazing abuse English public schools and the American Marines have often justified as a version of the life to come? Or is this a behavioural sink, an entire species turned aberrational, in Professor Anderson's words, 'an unusual population with a permanently disturbed behavioral repertoire'?

Anderson and his co-researchers have found some disturbing evidence that this is a kind of hell, a version of life gone
wrong. In a 2011 paper by Martina Müller and others, they
published 'the first evidence in a nonhuman [society] of socially
transmitted maltreatment directed towards unrelated young'.
It seems that Nazca booby society is a victim of the cycle of
violence so often encountered in the human world.

It begins with the murder of the sibling. If a bird has killed
a sibling, it is more likely to become a Nav than a bird from a
nest in which there was no other egg or in which the other egg
didn't hatch. Part of the answer is in that first life-shaping
experience. The first thing the Nazca booby knows is fighting
its sibling to the death. During that experience, testosterone
surges through its body and it becomes the governing experience for the rest of its life, what Martina Müller calls 'an
organizing agent' for adult abuse.

But there is more to it than that, as birds that had not killed
their siblings also turned into Navs. Something else soon
became clear: exposure to Nav behaviour made it likely that a
bird would later behave as a Nav. Abuse created abusers. The
cycle of violence would ripple on through the generations like
a curse. The whole population of Nazca boobies were destined
to suffer cruelty and abuse as nestlings until the end of time.

The Navs were no worse off physically than any other bird.
It was purely social learning that changed them and trapped
them. 'The siblicide experience is ephemeral but intense,
because it is a fight to the death,' Müller and Anderson wrote,
but 'a bird's history as a target of Navs proved to be a much
stronger predictor of its adult behavior.' They suggest that the
mechanism for this transmission of cruelty is 'socially induced
endocrine perturbations', or a permanent distortion of the
hormone system of the bird brought on by the terror of their
experiences as a nestling. Corticosterone, the stress hormone

present in all birds, of which the levels are reliably increased when the bird is under pressure of any kind, is running at high levels in the veins of every Nazca booby, driving them to a life of fear and violence, of which the only comparable example in nature is our own.

* * *

Approach Bass Rock, a tall mole-like lump of basalt off the east coast of Scotland, the biggest gannet colony in the world, and it is difficult to grasp quite what you are looking at. I have seen it from 30,000 feet when flying south from Aberdeen to London and thought a long plume of sunlit smoke was blowing away downwind from the summit. On another day in a boat, 10 miles away, coming from the mainland, there seemed to be a haze over the island, like moths above a summer meadow. Beneath it, the meadow itself was white, a summer snowcap, which from those miles away looked as if it might be guano. But smoke, haze, moths, snow, meadow and guano were all one thing: the white plumage of living birds. About 150,000 gannets breed on the rock, so many that there is now no room. Puffins that were once here have been driven out and apart from the gannets only a few shags, fulmars, kittiwakes and guillemots remain. Young gannets have been making their nests in the last couple of years down in the inter-tidal zone, so that the highest of high tides lap and flood where they are trying to raise their young.

Come nearer and the gentleness evaporates. The air is pulsating with birds, a reverberating repercussion of bird life, circling, many of them two-, three- or four-year-olds, with flecks and splashes of black in their plumage, populating the sky like the blizzard in a shaken glass souvenir, but constant, a churning flow and current of them, and endless like the creatures of a dream. It is screensaver nature, unbelievable in the continuity of it, looking as if they are not individual but on a loop, a universe of feathered things.

Tens of thousands of birds come to land among the others, some with feathers for their nests, carried in their serrated bills as if they were buccaneers with knives in their teeth, others roaring their identity and the news of their arrival, in a unbroken wave of roughened huzzahs, a regiment of Cossacks cheering and toasting their colonel and themselves, '*Ura, Ura, Ura.*' It is like a crowd talking loudly around you, a battlefield gone hoarse. There is no ease or calm in it but a bellow in the way an ocean beach can bellow at you, the unbroken sound of one voice and one booming trumpeted thunder call after another, more than 100,000 birds contributing to it, each ripple of the wave folding over and on top of the one before.

The North Atlantic's supreme ocean hunter

The hard black basalt of Bass Rock – it is a 'snout in the sea' as Seamus Heaney called Ailsa Craig, a 7-acre plug of magma in the neck of a long-gone volcano – is a place of unremitting and terrifying seriousness. There is no harlequinade here as there is in a puffinery, but a constant drive towards survival. When William Harvey came here with Charles I in 1641, he said the '*clangor & strepitus*', the roar and din, were too much for anyone to hear the person they were talking to. Go there today and you find a place of rage and desperation. Energy runs like whips through it. Anxiety, intensity and wariness is in every corner. Walk through the gannets and they slash at your calves. Stare at them and they stare back with a look beyond fear or hate, a gaze from machine eyes as heartless as a bank of cameras.

Gannets now breed in England no further south than the Yorkshire coast, but they fish off Norfolk and in the past were often caught in the herring nets. 'I met with one kild by a greyhound neere swaffam,' Sir Thomas Browne the great seventeenth-century Norwich doctor-scientist wrote, 'another in marshland knockd downe by a sheppheard while it fought & could not bee forced to take wing. another entangled in an herring net wch being taken aliue was fed with herrings for a while.' Browne looked hard into its 'wild blewe eye' and gave it the best (if inaccurate) name anyone has ever thought up for a gannet: *Larus maior ferox leucophaeopterus*, the great, white, dark-winged savage gull.

Gannet Ferox: that is the heart of this animal. With Maggie Sheddan, the guide for visitors on the Bass, one summer's day, filled with the ammoniac dust from the gannets' disintegrated faeces, I watch them around us. The snow slopes go on and on. In many places, the birds are so thick and so defensive that the colony is impenetrable. This is not a question of the conservationist's 'may not'; we cannot: it would hurt too much, even with the defensive plastic boards and leg protectors Maggie has given us.

The gannetry is a marketplace, a trading floor, a nursery, a battery-chicken plant, a fighting ground. A young bird, just short of fledging, with its black and white plumage still tufted in down, loses sight of where its own nest is and goes wandering through the world of its neighbours. At every turn, the built-up nests are set a careful beak-thrust apart so that the pattern of them is geometrical across the colony. The arrangement is like the hexagonal cells of a honeycomb, but here it is not the equal pressures of neighbouring bee bodies in the wax that has made the pattern, but the reaching-ring of those scalpel beaks. Each parent and child sit together in a state of armed defensiveness, their empire shrunk to an area no bigger than a small round table in an Italian café, 2 feet in diameter. Within it is safety, beyond it a thicket of knives. The slashing beaks are the barbed wire, the rim of each enclosure touching the neighbours' enclosures on all sides. The shallow valleys between the nests are what Berliners used to call the open ground alongside the Wall, the *Todesstreifen*, the death strip. Desperately the squealing squab jerks through this geography of pain, further and further from safety, wandering lost, stabbed in the neck and head at each new corner, scrambling for release but only delivering itself deeper into the killing zone. It is a scene of instinctive cruelty and destruction. Maggie looks on with a philosophical eye. 'We often try to do something for them,' she says, 'but the danger is we create chain reactions of displaced chicks and yet more damage. They almost always have to be left.' No seabird has a grave; most die at sea; this one, like many others, will rot on the Bass.

Why this ferocity? Why this competitiveness? Not all seabird colonies are such monuments to unkindness. Some Darwinian principle must be at work: kill the thing that is nearest to you, that is so near that it is in danger of draining the oxygen from the air around you, stealing the fish from your sea. Destroy the thing whose equipment and training and manner of being is so like your own that it can only be your darkest enemy.

Cooperation with it, even tolerance of it, would be a form of self-harm. To give them anything would be to deny yourself. And so this stinking, death-encrusted, vivid, howling all-life gathering of gannets and their offspring – so shockingly intense that it is in a category beyond the beautiful – is no colony, no monument to togetherness, but a giant clutch of deeply indi-vidualized beings as much each other's allies as arrows in a quiver, or athletes on a start line, or the axes bound in a Roman *fasces*. I have never witnessed life less accommodating. Bass Rock is a magnificent demonstration of existence clenched in hate.

That is, of course, not true. The gannets do an astonishingly good job of protecting and feeding their young. They clasp the single egg within the warm webs of their giant grey feet, a Madonna gesture holding the unseen life in their hands, trans-mitting warmth through the shell to the embryo inside, until the day comes, two days or so before the chick breaks the shell, when from inside the egg it begins to call for its mother and father, hearing them when still shut in its first shell-world. Later, as the chicks grow through the summer, looking at times like solemn-beaked judges in their full-bottomed wigs, the parents nurture them, the parents sitting next to their ungainly progeny, and raising a successful young bird in three nests out of every four.

In the early 1960s, Bryan Nelson was the first person to look long and hard at the gannet. He had been studying black-birds in the woods outside Oxford but, longing for the wild-ness and roar of seabird islands, he went to see Niko Tinbergen, then established with his gaggle of doctoral and postdoctoral students – many later to be famous, Desmond Morris, Mike Cullen, Hans Kruuk and Richard Dawkins among them – in a collection of hen huts on the roof of the Oxford University zoology department. Tinbergen dispatched Nelson to the Bass,

Gannet chick, dressed in a judge's wig

where he and his wife spent the next three years (honeymoon included) examining gannets, occasionally returning to Oxford for a wash and a debrief.

The Tinbergen method – long, careful observation of bird habits; constant questioning as to the motive of those behaviours; looking for a relationship between behaviour, body-form and the principles of natural selection – drove the research, above all in quest of an answer to the one overriding question: why are gannets so extraordinarily aggressive?

Nelson set up his hide on the outer rim of the gannet crowd and began to watch. Between the nests, he reported, 'lies a glutinous black ooze of mud, decayed seaweed, ordure and spilt fish. It may be more than ankle deep and the stench rises when the crust has been punctured.' When two young male gannets decide to set up a territory early in the year, they inevitably scrabble and squabble in this foetid ooze and 'erstwhile immaculate males became plastered in mud, pockmarked with stab wounds on the head, torn and bleeding about the face, with pale blue eyes crackling electrically from the mud mask'. The fights could last for half an hour, sometimes for two hours, locking bills, wings threshing the ground, the neck strength of the birds, needed for plunge-diving, allowing them to fight like prize wrestlers, lacerating the skin on their faces and inside the mouth, sometimes destroying an eye. These passionate and ferocious encounters were so exhausting that after a fight the bird might sleep continuously for two or three days.

It is not for want of room. 'A lovely windswept spacious site, utterly uncontested, is no good to a gannet if it is several yards from other nesting gannets and the space between is suitable for breeding on.' Only the places immediately next to existing nest sites will do, and they will do to the extent that they will be defended and fought over to the death. There are plenty of cliffs and islands around the North Atlantic that would be suitable for gannets but they don't choose them. Only the vast agglomerations in which competition is bitter and fighting is ferocious are satisfactory as places in which to breed.

Why? It is not that the gannetries are particularly near sources of food. Gannets regularly fly on fishing trips of 350 miles or more away from the nesting site. Perhaps a great collection of gannets is better protected from predators than a solitary nest would be. Very few birds can prey on a gannet. In Shetland, great skuas occasionally kill adult gannets and in Norway sea eagles take them. Colonies there have declined and even been wiped out through sea eagles killing and disturbing the birds. But those are relatively rare conditions and, apart from that, there is only one habitual killer of gannets: man.

Nelson established that chicks hatched in the centre of a colony were in better shape at the end of the summer than those from the edge, and so revealed an unbending gannet rule: breed as near the middle as you can. And the connection between that need to be central and the need to be aggressive was clear. If competition to be central is that fierce, only if you could keep your neighbour-rivals at bay would your chicks thrive.

Rivalry was one thing but breeding was another. And among Nelson's most important discoveries was his recognition of the fascinatingly intimate relationship of love and hate in the gannet psyche. How, he asked, have the gannets managed to steer their ferocity into the necessary, fruitful paths? During sex, the male always bites the nape of the female's neck, and quite often attacks her when she returns to the nest. Ferocity is so near a gannet's heart that it has invaded his most intimate moments of love and affection.

Nelson began to work out the ways in which the birds have evolved two rituals, one of affection (a word Nelson was happy to endorse) and one of effective parenthood, to allow them to breed. The love ritual of the gannet, often repeated, is a kind of mutual fencing, male and female sparring with their most dangerous weapons, striking and knocking against each other, bills held high, breasts touching, a kind of play fight which is the essence of intimacy, constantly confirming the closeness of the bond. And when one or other wants to leave, they raise their bill to the heavens, sky-pointing, a non-assertion of self, looking to the other bird to see that it will *not* sky-point and will *not* leave, so that the egg and the chick will always have a parent with it on the nest. Sky-pointing has this one message for the mate: I am *not* being aggressive; I am relying on you to do your part.

Mutual fencing and sky-pointing are behavioural twins: both are highly demonstrative and both often repeated, but one is play-fighting and the other is not-fighting. One says I

Gannets use their beaks – the most powerful weapons they have – to
demonstrate affection

love you, the other says I need you. In both, the most terrifying instrument the gannet possesses is used to confirm an
attachment to the other. The only thing that can make that
confirmation is the weapon of total violence. No gannet can
survive without both ferocity and intimacy. The function of
gannet social life is to allow the fierce to love.

For all this ingenious and revelatory natural history, Nelson,
despite decades of looking and thinking, never established why
living in such a dangerous colony was worth it. He was sure
only of the secondary sequence: chicks in the centre of the
colony do well; gannets know this and so have to compete for
the good spots; to compete they have to be fierce, or to put it
another way, only the birds with the fierce gene will thrive and
so in time that gene will spread through the whole population.

He also established that it wasn't the centrality of the nest that guaranteed a good strong chick, only that most central nests were occupied by older and more experienced parents. If a young gannet male happened to get himself a central nest, defend it and invite a female to join him, their chick did no better than chicks from the outer edges of the colony. It wasn't centrality that made good chicks, but good parenting. That had been difficult to see at first only because old birds usually occupied central nests. So the question took another step back: why is it a universal gannet principle that a central nest is a good idea? Why is crowdedness so good that it is worth the risk of exhaustion, blindness and death to achieve it?

Bryan Nelson's decades of dedicated labour were unable to find an answer. Only recently has a combination of evolutionary theory and technology started to provide one. In the early summers of 2010 and 2011, when the gannets were feeding young chicks, a team of bird scientists from 14 universities and NGOs put a variety of trackers on 184 of the birds from twelve colonies dotted around the shores of Britain, Ireland and northern France. They wanted to see where they were fishing.

The results confirmed something that had only been suspected before. It had long been assumed that the fishing grounds of nearby colonies would overlap. Why shouldn't they? What could prevent gannets from different colonies going after the same fish? But that wasn't the picture revealed by this first comprehensive study. It was clear that the gannets had divided up the sea. Each colony had its own fishing grounds which birds from other colonies almost never entered. Where gannetries were near each other, such as Little Skellig and Bull Rock in south-west Ireland, or Grassholm on the east side of the Irish Sea and Great Saltee on the west, the gannets would set off to fish in different directions. Little Skellig birds almost

Satellite tracking of gannets around the coast of the British Isles
revealed that individual gannetries had their own stretch of sea in
which to fish, with almost no overlap between different colonies

always went north, Bull Rock birds almost entirely south.
Great Saltee gannets stuck to the Irish coast, Grassholm birds
ranged widely from the Western Approaches to the coast of
Cornwall and to the Gower Peninsula. The two burning questions these results stimulated were: why did gannets fish in

this pattern? And how did they divide up the sea into these mutually exclusive, colony-specific home ranges?

The new hypothesis goes as follows: when two gannetries that are near each other start to grow by natural increase, the area of sea in which the birds fish is likely to grow because they are depleting the nearby resources. There inevitably comes a point when the home ranges of the two colonies overlap. To begin with, birds might fish together, but competition will be hot and, increasingly, they will start to fish further away where the competition is less.

But they need to know where that is. Random searching for fish, or even remembering where they have fished before, is unlikely to steer them away from that competition. Only by observing gannets from the same colony, particularly those gannets which are doing well in raising good strong chicks, watching where they are going and watching where they are returning from, will lead the young, inexperienced gannets to that part of the sea where they are likely to find fish not already fished out by the neighbouring gannetry.

This fusing of 'public information' and memory quickly establishes segregated ranges for birds from neighbouring colonies. Information is channelled from colony to bird quite unintentionally. The individual birds are anxious receivers of information but not deliberate transmitters of it. But the effect is for each colony to develop a set of habits, a fishing pattern, a way of doing things which is unique to that colony, passed down across the generations, creating what is in effect a culture, a pattern of understanding and a way of life, tied to its own geography, unique to that gathering of gannets. Memes, or cultural clusters of knowledge and skills, are inherited across the generations.

Gannets have cultures by which they live and make decisions. Their nosiness about their neighbours' affairs is the

Young gannets only survive by learning from their elders

source and method of that culture. The colony is an information centre, not a battleground but a university of life and
survival. The unending stream of young birds, winging round
and round the colony day after day for years before they first
attempt to establish a nest site and a pair-bond with a mate, are
all learning. Look at them and you will see them staring down
intently at the scene below. The assumption is that they are
looking at how successful breeding birds are managing their
lives: at the habits of older paired birds and their chicks; at the

shape and behaviour of arrivers and leavers; at the condition and welfare of chicks; at the excellence of life in the colony centre; at its desperation on the margins. This hunger for information explains the intensity of violence in the life of the gannet. Unless it can get close to that source of information, it will not thrive. The bill-slashing tussle of the birds is the ferocious competition between applicants to the information core of the colony, that piercing eye and that serrated bill the most powerful demonstration imaginable that seabirds live by learning from their elders.

Many vertebrates, including the dinosaurs, have nested in colonies, and nearly all the seabirds do. The very few seabirds that nest solitarily, like the marbled murrelets of North America which lay their eggs high on the mossy branches of old pine trees up to 50 miles inland, are clearly embarked on a very different and relatively recent evolutionary path. Colonies work for most seabirds even though the costs are high: parasites and diseases, males cuckolding the neighbours, competition for food and mates, cannibalism and infanticide. The benefits must be good to make it worth it.

How do seabirds make that calculation? It may, in fact, be simple enough. The presumption in a bird's genes may be: I have been born here, I am alive and thriving and so I will return to the place I was born because, if it could make me, it could make others like me. Clustering makes sense. Many birds return to breed where they were born and, for those that breed away, the second calculation may be just as simple: that other colony is clearly producing viable chicks. It looks right and smells right. Other birds are fighting to be there and so it would be a good bet for me to be there too. The presence, apparent strength, beauty and breeding success of birds in a colony will be the criteria by which any bird decides to join. That thought process will generate the urge to cluster, to get as close as possible to the source of well-being. The colony itself is a barometer both of the environment around it and of the gene pool present in it.

Density is beautiful for a gannet. Roaring is probably beautiful. Fighting is beautiful. Fat chicks will certainly be beautiful. That careful investigation of others like them is one reason why seabirds are not spread in a thin skin around the edges of the ocean. Add to it the fact that seabirds also need breeding sites that are secure from ground-based predators – usually offshore islands with sheer cliffs. Those opportunities may be limited, and in that case individuals will be forced to breed in crowded colonies, the only safe places they can find. All these factors contribute to the phenomenon we now see. Densely packed, deeply secure information exchanges on the shores of richly stocked seas: that is the seabird's idea of the beautiful place.

The tightly linked triad at the heart of seabird life – ferocity, coloniality and information exchange – are all dependent on a fourth element: it is not always clear exactly where the fish are to be found. Fish are unlikely always to be where you hope they are. They are not like the seeds of grasses or fruit hanging off nearby trees. Up-to-date information on the whereabouts of food is vitally necessary for birds. Colonies provide that information for seabirds, but landbirds do not need it. They know where the food is – and the effect is clear: 98 per cent of all seabirds nest in colonies, but only 16 per cent of landbirds do. Landbirds are better off defending an individual territory – you can't defend a shoal of sandeels – and so for them solitary nesting in individual territories makes more sense.

Solitary feeders would soon die looking for the fish at sea. Only the colony's shared information-gathering and communicative system can deliver up the riches that cannot be seen. Was it life at sea, and the sheer invisibility of the prey, that

forced species to become colonial? Or was it being colonial
that allowed certain species to make use of an environment in
which food is too patchy and unpredictable to be any good to
birds that breed alone? There is no answering that question
but it may be that the ancestors of the seabirds abandoned
territoriality before they arrived at the sea and, like most of
the swallows and the swifts, started to nest together, feeding
as they flew on clouds of insects that are just as unpredictable,
in both time and space, as any fish. That thought brings on a
sudden sense of time-vertigo: when watching the swifts flicking and screeching above the tiled roofs on any summer
evening in Tuscany or Portugal, in Somerset or Crete, is it
possible you are looking at birds whose descendants will, one
long-distant day, also respond to the summons and challenge
of life at sea?

* * *

No gannets breed on the Shiants. For a while in the 1980s, a
single solitary bird used to perch for a few summers on the
northern cliffs among the guillemots and kittiwakes, but no
mate ever came to join it and the solitary Shiant gannet even-
tually disappeared. But from those cliffs, out in the wide
stretches of the Minch extending on all sides, gannets often
come to fish. Beyond the limit at which eyes can really see,
visible only with binoculars, and even then to be distinguished
from the white sky behind them only now and then, coming
and going, seen and then unseen, to the point where you can't
be sure that it is actually there, a fleet of gannets plunges for
the kill. Even 10 miles off, they can flicker in the sunlight as
they cruise and turn, lightless and then lit, at the limits of the
graspable. No one cannot watch them. The poet Norman
MacCaig, seeing them 'fall like heads of tridents' in the waters

off the sandy beaches of Assynt on the Scottish mainland, recognized that repeatedly, in that moment of their falling, the sudden turn and the accelerations of gravity, 'Space opens and from the heart of the matter/sheds a descending grace' – as concentrated a description of what it is like to be in the presence of hunting gannets as has ever been made.

It is the most theatrical sight of life at sea, the heaviest plunge-divers of the Atlantic going for prey. Gannets can see each other fishing from nearly 20 miles away; it is one of the reasons they are white, the diving bodies communicating the presence of fish across a full circle around them, 1,200 square miles of sea. Gannets have been tracked flying high so that they can lock on to this precious spectacle from beyond the curvature of the earth. They have been known to dive into nets full of fish, into the scuppers of fishing boats and into their holds, and once straight into a fish-curing shed in Penzance. They will eat anything: a gannet can swallow four large mackerel in succession, ten herring or a cod thicker than a man's forearm. A 2-foot-long brass welding rod and a long loop of iron wire have both been taken from gannets which had been found dying as a result of swallowing them.

It is the sight close up that stays in the memory: the cruising survey, the sudden drop, the turn in mid-air, the hunter-fighter's variable geometry, arrowing for the sea, wings folding back and then, within a micro-moment of entering, fully extending behind it so that the body offers no resistance to the water.

But it does and you hear it, the *wheeumph* of a muffled detonation as the 7-pound gannet hits the surface at 80 feet a second, dropping from up to 90 feet at something like 50 miles an hour. The whole of the bird's face, neck, stomach and back is cushioned in an interconnected mattress of air-sacs to absorb the impact, and from which the air is immediately expelled as it penetrates the water so that the gannet drives 10 or 15 feet into the shoal. Its nostrils have migrated to inside the closed beak so that no water is driven up them on the plunge. And its eyes face forward to give it the binocular vision needed to calculate the distance at which the wings must fold.

'A bird is a wing guided by an eye,' the great French ophthalmologist André Rochon-Duvigneaud wrote famously, and the gannet eye remains one of the miracles of seabird life. If you look at a selection of bird eyes in a taxidermist's cabinet, the gannet has the smallest pupil and the palest, clearest iris, a needle of darkness in an aquamarine desert, the maximum cold, the minimum generosity. In air, it is the cornea, the surface of the eye, that fixes the image on the retina, with which it sees the flicker of herring or mackerel or sandeel beneath the sea. But in water the cornea loses that ability and it is the lens which focuses the image. The challenge for the gannet is the rapid transformation from aerial spotting to aquatic hunt, and a fascinating investigation of Australasian gannets by the Sydney biologist Gabriel Machovsky-Capuska, filming them under water and then analysing the video frame by frame, has shown how within a twelfth to an eighth of a second after entering the water, the bird has adapted its optics, squeezing and distorting the lenses of its eyes so that it can see clearly. At the same moment, the bird opens wide its irises – submarine gannets are black- and not blue-eyed – a trade-off between lower resolution, with a much shallower depth of field, in return for a brighter image of the fish in the shoal.

The plunge phase works by surprise. The turbulence and bubbles created as the gannet enters the water probably make it difficult to see anything at all; the gannet will be travelling at a speed equivalent to the fastest a fish can swim; and with its wide-open, unfocusing eyes will grab at anything within reach. In a big feeding frenzy, on a shoal which many gannets are diving for, birds can be killed in underwater collisions as they dive blind one on top of another. If the plunge delivers a fish, the gannet will then bob to the surface under its own buoyancy, completing a V-shaped dive. If not, it then enters the second phase, now a U-shaped dive, swimming with wings, tail and feet and porpoising like a swimmer doing the butterfly stroke, down to as much as 90 feet under water, actively searching out prey. It may be that in the plunge it sees the shoal, and in the U-shaped dive, with its eyes now seeing clearly in the water, it sees individual fish.

Fishermen have long watched the gannets and understood that they are part of the life-web in the sea. Donald 'Dolishan' Macleod, fisherman in Marvig, on the eastern shore of the Isle of Lewis, remembers how sometimes the herring were so thick that 'you could see black sea, big black splodges of herring, they were that plentiful then and at night, with the phosphorus in the water you could see the herring then, you see them moving'. The birds were surefire guides.

> You would see the gannets hovering above and you see them diving, just for the mackerel. The mackerel was always why they were up. But if you see them going straight down, and staying down for a while, they were down for the herring. After they had been down for a while, you would chase them for a while, see if it spews something out. If it spews herring out … they were that full they couldn't fly. If they would start spewing out herring, there must be herring there.

Gannets live by watching in the same way fishermen have always watched them. They know their world by familiar waymarks, they fish where they and their forebears have fished, they must respond to the daily difficulties and contingencies of

weather and sea, but underlying those innumerable daily decisions is a fixed mental map of what is possible and what has
worked before. It may be that the mentalities of traditional
human fishing communities are as near to the mentality of the
gannet colony as we might ever get.

The poet Katrina Porteous has written about fishermen on
the coast of Northumberland where she lives. They have long
understood the principles of the sustainable catch, and have
long been so familiar with the local sea floor that they could
walk it in their dreams. For them it was a question of embed-
ding their expertise and understanding in the natural cycles on
which they relied.

Before echo-sounders or fish-finders, they had to know
their place in the sea and follow its paths, by the landmarks
that had guided their ancestors over the generations. In her
poem, 'The Marks T' Gan By', she tells how the fisherman
Charlie Douglas remembered for her the contours of that
half-forgotten world:

> His fingers twisted round the slippery twine
> In the stove's faint firelight. It was getting dark.
> 'Them days,' he said, 'w' hetti gan b' marks.
> 'Staggart, the Fairen Hoose; Hebron, Beadlin Trees …'
> Thus he began the ancient litany
> Of names, half-vanished, beautiful to hear:
> 'Ga'n roond the Point, keep Bamburgh Castle clear
> The Black Rock, mind. Off Newton, steer until
> Ye've Staggart level the Nick a the Broad Mill.'

The string he twists in his hand is the string of inheritance and
continuity, a chain connecting him to the past and the
natural-cultural world which he knows in his bones is the good
one.

The sea's the boss. Me fatther telled me so.
'Them marks,' he said; 'he handed aal them down
Like right an' wrang.'

Gannets think and behave in just that way. In June and July
2010, eighteen gannets from Grassholm in Pembrokeshire and
another thirteen from Rouzic in Brittany were fitted with GPS
loggers. The 30,000 miles of gannet-flight recorded over about
four weeks revealed that both the French and Welsh gannets
consistently flew out to fish along the same paths and to the
same destinations, not in gaggles but in individually distinct
patterns, back and forth to their own favourite fishing spot,
some to where zooplankton had thickened, still others to
places where the sea-surface temperature was higher. But
those differences remained consistent. The different birds
were specializing, catching different prey, fishing in different
places, dividing up the resources of the sea between France
and the coasts of western Britain. One Rouzic bird went back
and forth to the Channel Islands. Another loved the Lizard. A
third consistently visited South Devon. From Grassholm, one
was always returning to Cornwall, another going east towards
the Bristol Channel, a third north into the Irish Sea and a
fourth repeatedly exploring the Celtic Sea to the west. The
gannetries were thick with individuality. The colonies were
bird societies alive with variety and character, with individuals
knowing where they were going and what they wanted.

Further studies on the same birds followed. Samantha
Patrick of Liverpool University and her colleagues from
Exeter University have mapped the behaviour of some of these
gannets in relation to fishing boats in the Celtic Sea. She was
able to show that individual gannets had developed very different ways of being. Some birds almost never hung out with the
boats, some did almost nothing but and some varied their

methods on different trips, sometimes with fishing boats, sometimes on their own. These differences last from one year to the next and are the product of what Samantha Patrick called 'personality differences'. She has suggested that the different opportunities available in the sea – following a boat, finding an ocean front, detecting warm or plankton-rich patches of sea – 'may lead to the emergence of such personalities'.

It is the deepest possible insight. The sea itself is a gathering of tens of millions of different niches, into every one of which gannet life will flow. The sea does not make 'the gannet'. It makes hundreds of thousands of gannets, each an individual. Gannets are booming at the moment in the North Atlantic, individual birds making individual choices, conducting individual struggles and achieving individual triumphs. It is not the *Umwelt* of the gannet we need to understand but the *Umwelten* of all the gannets there are.

9

Great Auk and its Cousin Razorbill

From all over the big lichened boulder colony that looks out on to the bay at the centre of the Shiants, razorbills are summoning their chicks. It is a warm July afternoon and the tide is easing slowly between the islands. Flies buzz in your ear and you might almost be tempted to swim. The chicks are about three weeks old, half the height and a third of the volume of their parents, still without the big razor-bill, but feathered in the same black and white as the grown-up birds. They look overdressed, readied for their first adult experience. At each opportunity the young ones fiercely and frantically beat those stumpy, undeveloped wings, stretching muscle and tendon. No primary feathers have yet emerged and the wing area is still tiny, but this is an awakening of life beyond the split or crack in the rocks which has served as their half-good nest until now. No sign of the grey fluff of the newborn survives on them.

In the daytime, you scarcely see them. They are still hidden within the crevices, but their fathers – and it is always a father's duty among the razorbills – are calling for them to leave. The economy of parenthood has established that, for this species, taking them out to sea when they are small is the most effective strategy. It reduces the costs of bringing fish back to the chick; and it probably reduces the threat of predation from other occupants of the islands. The sea is both food and safety. Unlike the puffin, whose chicks leave the burrow when they're much older but are alone from the beginning, the razorbill father remains with his chick out at sea for several weeks until it can fend for itself.

And so on summer afternoons and evenings the boulders rattle with the paternal cry: *rrraaar rrrrarr*, a miniature roar-cry, leonine, Lilliputian, peremptory, instructive, deeply affecting. The razorbill has his head held high in the air, flashing his wings as he calls, his big yellow gape open to the sky. From hundreds of yards away you hear them, the extraordinary resonance built into these little bodies, so that the sound drums through them. Their whole being is a voicebox. The fledglings reply, shrill, high and bright from within their crevices, a pure note, so that it is a duet between rasping father and peeping child, *rrrrar-peep*, *rrrrar-peep*, dialoguing across the generations.

Shiant razorbills

This is the anxious transitional moment in which the whole year's investment is at stake. The razorbill beside me is ducking his head down into his neck, jumping backwards and forwards outside the nest. Come on, come on. His head flicks back and up, *raaar ra raaaar*, a call to action. It is about 10 past 5 in the afternoon.

He stands there and calls for his child, leaping from one boulder to another, fluttering his wings for stability each time he lands. A shag stands quietly beside and above him, looking down at this drama as if disengaged or even bemused by the razorbill's antics. It is the way non-parents look at parents trying to cope with a recalcitrant child at a party: faintly predatory, faintly indifferent.

The razorbill is continuing with his call, the note roughening, tooth-grating, as uneasy as the rubbing of two emery boards. The shag stands there as some kind of beadle at a funeral, gorgeous, dark, disconnected from the ordinary, the embodiment of fate. I realize what is happening.

The razorbill dad moves towards the hole from which the hidden chick is peeping. Nothing is more insistent in this tiny drama than the parent's lack of control. He does not have the capacity or the instruments to persuade his chick to come down. Surely this would be better at night, which is when most of the razorbills perform the task? Perhaps this is a young father, unpractised in these things. His call is now shrinking: *dar dar dar dar dar*, his neck further and further back, his anxiety tangibly increasing as he lives through his frustration.

The neck of the shag is looking purple in the light off the sea. In the heat of the afternoon the bird is fluttering the skin under its bill to cool. The razorbill is still calling. By about 5.25, he seems exhausted, his cry muted and his head barely up. Come on, come on, the genetic imperative. But how is the chick going to make its way past that dark predatory figure above him? The father's call is fading and the chick is no longer replying. I have seen others this morning out at sea, little pairs of companions, one teaching, one following. But these two are late, lonely and exposed.

By 5.32 the unseen chick is squeaking like the unoiled wheel in a pulley. The dad now makes another effort, *wow o wo woow-owow*, and another razorbill far away, a quarter of a mile down the colony, does the same. The shag looks on in silence.

Razorbill chick on the point of going to sea

It feels like an impasse. Time goes by and it is 5.40 when suddenly the chick appears out in the open on the lip of the rock crevice, coming towards its father. There is a rush and the shag blunder-flusters towards it, drives the adult razorbill away, and for a terrifying moment the razorbill chick stands gawping and incompetent in front of that shag bill, the sheer inability of the razorbill father to do anything for it ballooning around us. The shag jumps down into the hollow to which the chick has retreated. Its little body is within reach of that shag mouth, experiencing what can only be horror as the shag comes for him, but the chick now stumbles and falls down and away, too far into the hollow for the shag to reach it and the big dark-footed shambling bird shuffles off away to the rock on which it has sat all afternoon, where it then preens and nibbles at the feathers on its wings and back.

The father razorbill joins the chick down in the crevice. Both now are beyond the reach of any enemy. The father reappears on the lip of the crevice and calls again, harder and tighter than before. Come this way, this way, this way, but the chick makes no sound in return.

After a few minutes, I go over to where the razorbill is standing. He flies off out to sea. I can look down into the gap between the boulders and with my phone shine a light there to the little alert eye of the chick, a button of sensibility deep down in the dark, the light reflecting off its perfect feathers. It is too far for me to reach, too far for the shag, too far for the father and too far and too vertical for the chick ever to emerge.

* * *

Some part of me thinks of these animals as blades of grass, or as the stalks of a reed whose giant shared rhizome is far below in the mud: the individual stem has no significance. What matters is the population, the genetic mass of which it is a part. What is this chick but a surface phenomenon, one of the foam-bells on the currents of being? That is how this system works and that is what the half-hour drama on the boulders signifies. The birds must play percentages, and this is a game these razorbills have lost. The annual razorbill success rate varies from colony to colony and year to year but on average it is about 0.7 fledged chicks for each pair. I have just witnessed the fate of the other 0.3 unfolding in front of me. Of those fledged chicks, only one in five or six will ever breed and so of all razorbill eggs ever laid, something like twelve or thirteen in every hundred will produce a breeding adult. The nutrients embedded in the other eighty-seven will in time drain back to the sea where the creatures on which these birds or their descendants feed will once again recycle them. That is the essence of the colony; cyclicality is its continuity; and without death there can be no life.

That micro-cycle, from the white-and-black-splotched egg laid two months ago, to the chick hatched five weeks later and on to this little terminus on a sunny afternoon, has its larger equivalent which takes more accommodating. 'No fact in the long history of the world is so startling', Darwin wrote when off the coast of Patagonia on the *Beagle*, confronted with the fossil evidence of the South American past, 'as the wide and repeated extermination of its inhabitants.' Since the first birds flew 140 million years ago, about 150,000 species of them have evolved. Of those vast flocks and armies of unknown birds, some 140,000 species are now extinct. It is not only the chicks that get caught down in the darkened crevices between the boulders. Extinction is an aspect of how things are.

The poet and librettist Melanie Challenger has described how her grandmother 'keened over the blunted places of her childhood'. Disappearance and the diminishing of treasured things has the same kind of fascination for us as that black tongue of water at the centre of a river plunging over a fall. It is the place into which all energy flows and vanishes. 'The idea of loss is riveting to the imagination,' Challenger has written, 'like shrapnel lodging in the mind.' She revived the extraordinary Anglo-Saxon word *dustsceawung*, meaning 'the fascination experienced by someone looking at a ruin, a kind of daydream of dust, pondering that which has been lost: dust-*seeing*, dust-*chewing*, dust-*cheering*. The daydream of a mind strung between past and present.'

Everything in a great bird colony is the daydream of a mind strung between past and present. Continuity and extinction orbit each other there, twin stars, one light, one dark. A small change in the mathematics, a slight overtipping in favour of the predator, and that razorbill cycle would move from the wonderful toy-lion roaring of the fathers and their peeping chicks to absence and silence. Extinction is an ever-present possibility even as their fat, swollen life is blazing around you. It is also, given the history and time-length of a species's existence, a certainty. The moment will undoubtedly come when the only razorbills that survive are fossils. We will probably not be here to witness it.

The death of a razorbill is made more poignant because they are the representatives on earth of the largest seabird ever to have lived in the northern hemisphere and its most famous extinction, their most magnificent cousin, the Grand Pingouin, the *Pinguinus impennis*, the flightless auk, the *falcóg mhur*, the *arponaz* in Basque, the *geirfugl* in Germanic languages or since the eighteenth century, when the Welsh naturalist Thomas Pennant gave them this name, the great auk.

Watch the razorbills and you are coming as near to a living great auk as you ever will. When in the nineteenth century stuffed great auks or their skins were dispatched for sale to the gentleman collectors and museums of Europe and North America, a couple of spare dead razorbills were often sent with them in case the great bird needed repair. In plumage, in wing-form, in bill-shape and in egg-patterning, the razorbill perpetuates memories of the life of its great cousin. Like the razorbill, the great auk had a tendency to shudder and shake its head on the nest, a constant self-adjustment of body and

Edward Lear's depiction of the great auk from John Gould's *Birds of Europe*, 1837

plumage. The razorbill has a fine white line on its face, the great auk a big blodged white patch in front of the eye, plus the powerful, ridged bill, all of which suggest that like the razorbills they used their body and its markings to establish and confirm their pair-bonds, laying a single large egg most years and remaining monogamous for long lives. Only in size – one of the most elastic and adaptable of animal characteristics, quick to change over relatively few generations – does it differ. But, as with the blue whale, size was what drew the hunters to the great auk. Its sheer quantity of feathers, meat and oil made it their favoured prey, so that now the great auk stands for all endings, and all rememberings, the poster-bird for the species we have lost or destroyed.

The last great auk on St Kilda 'made a great noise shutting its mouth', the St Kildan Lachlan McKinnon remembered as an old man in the 1860s. It had been killed forty years before. Its cry, McKinnon said, was 'like that made by a gannet but

much louder'. He and his two friends grabbed it on one of the
St Kilda stacks. It was fast asleep. But once they had caught it,
'a storm arose and that fact together with the great size of the
bird and the noise it made with its bill caused them to think it
must be a witch … It was killed on the third day after it was
caught and they were beating the bird for an hour with two
large staffs before it was quite dead.'

That was probably the last British great auk. The last one on
the western side of the Atlantic had died before 1800. And the
last of all, probably, but with no certainty, was killed in Iceland
in 1844. Plans are now afoot to recreate the great auk, using a
combination of razorbill genes and its own, taken from one of
these preserved bodies and inserted into something like a
swan's egg which could accommodate the giant chick.

The American zooarchaeologist David Steadman, the great
investigator of the fate of extinct birds, says that 'We expect
extinction after people arrive on an island … Survival is the
exception.' Nor is it the Europeans who have solely been
responsible for this destruction. The voyages of the Polynesians
across the Pacific in their sailing canoes is one of the great
stories of collective human enterprise but it brought extinction with it. More than 2,000 bird species have been destroyed
by man in the Pacific, a fifth of all birds now in existence, most
of them dead before about AD 1300.

We are the holocaust, the destroyers of what we come to
live with, and not only because we hunt them. We bring with
us the other three horsemen of the apocalypse: habitat destruction, diseases which the indigenous animals have no defence
against, and the introduction of other predators, the snails,
rats and cats which eat their way into the lives of animals that
do not understand them or what they represent.

There were once flightless owls in Crete and a giant flightless swan in Sicily and Malta, but both were killed as soon as

people landed on those islands. The black Canary Islands oystercatcher, last seen for certain by lighthousemen on the Isla de Lobos north of Fuerteventura in the 1940s; four small island petrels, one from the Canaries, two from St Helena and a nocturnal one from Jamaica; three shearwaters, one from Majorca and two from the Canaries: all have disappeared from the Atlantic world.

The great auk remains the King of the Lost. It was the first 'penguin', maybe a Breton or Welsh name, which the French still use for the razorbill. 'Penguin' was only later transferred to the flightless birds of the Southern Ocean when early European sailors were reminded of the auks they knew at home. The last of the great auks were found and killed in the far north, in Newfoundland, Scotland and Iceland, but those were only the safest, most distant and residual refuges. They had once stomped and hunted across as much of the Atlantic as the penguins now cover in the Southern Ocean. Fossils have been found in Calabria in southern Italy, on the Atlantic coast of Morocco, in Gibraltar and in the Canaries. A whole landscape at least 500,000 years old has been uncovered in a quarry at Boxgrove in Sussex, and here, alongside eagle owls, ancient swans, geese, gannets, cormorants and razorbills, were great auks with the bones and stone axes of the men who had butchered them. Peat bogs in Denmark, islands in the Baltic, the foundation levels of Stone Age houses in the Scottish Hebrides and in Ireland, over forty places in Norway and Sweden, a fourteenth-century village in South Wales: all had their great auks. They were in west Greenland 4,000 years ago, in Labrador and Maine and perhaps in Florida, although those bones may have been taken there as food. The body of a man in one prehistoric grave in Newfoundland has been found covered in more than 150 great auk bills, perhaps the remains of the most astonishing seabird cloak ever made, clacking and rustling around the body of the ancient chieftain like great auk chain mail, a sheath of Atlantic bird life.

But these birds play a more heroic part in the history of human consciousness than as materials merely for dinner or display: they were the first to be depicted by man. The Grotte Cosquer, named after Henri Cosquer, the diver who discovered it in 1985, is in the dry, white limestone cliffs of Provence to the east of Marseilles. Twenty thousand years ago, when the last Ice Age was at its coldest, the sea level was 150 feet or more lower than today and the entrance of the cave would have looked out across a level coastline to islands offshore. Now that entrance is 115 feet below sea level and the only way in is with dive equipment. From the mouth a gallery slopes up for about a hundred yards until it surfaces in an enormous cavern. In here, with a few fires, using torches and flint tools, palaeolithic men painted the walls, in two periods, the first about 27,000 years ago, the second 8,000 years later. They painted many hands on the walls, both solid and in outline, several with apparently shortened fingers, and 177 animals, most of them the desired prey in a meat-loving world: horses, bison, aurochsen, ibex and red deer, both stags and hinds. There is a big cat and some chamois, and, almost uniquely among palaeolithic cave-paintings, sea creatures: seals, a human being with a seal's head, four fish and three great auks.

High on the wall of the cave, two of the auks face off towards each other, not side by side but confrontationally opposed, their little wings outstretched, the head of one held slightly back on its long thin neck, the other more insistent, perhaps with its mouth agape. Both have their feet firmly planted at the bottom of big strong curved bodies. The outline is unmistakable. Some flaking of the wall surface has degraded the picture a little, but this is clearly a scene from life, of creatures whose manners and ways of being had been noticed and absorbed.

Alfred Newton, the Victorian professor of comparative anatomy at Cambridge, after visiting Iceland in the hope of finding great auks still alive, relayed an account of how the auks had behaved when approached by Icelandic hunters. They 'showed not the slightest disposition to repel the invaders, but immediately ran along under the high cliff, their heads erect, their little wings somewhat extended. They uttered no cry of alarm, and moved, with their short steps, about as quickly as a man could walk.'

In Grotte Cosquer, from 19,000 years ago, that is exactly how they are shown: two of them with their wings outspread, a busy alertness in them. A third is settled with its wings folded a short way below, looking on as the confrontation occurs. The opposed pair at Cosquer, as the palaeontologist Francesco d'Errico has said, 'might be males in combat, the third the female they are disputing over. The dynamism of the pair and static appearance of the third lend credence to this interpretation. Battles between Great Auks had to be impressive because of the birds' size.'

Although all auks are on the whole monogamous and stay with the same partners for years, fights and squabbles between neighbours are a constant of colony life. Not only is this the first reliable depiction of a seabird; it is the first depiction of seabird behaviour, animated by territoriality, competition and sex. The walls of the cave are peppered with depictions of both male and female genitalia, making use of natural hollows and protuberances in the rock. In among that highly sexualized surface of the cave, fingers of what are clearly children's hands have scraped into the soft limy clay on the cave walls. This deeply hidden and powerful place, filled with the creatures of a shared and living imagination, was somewhere in which the meaning of sex and fertility was all important. Animal and human, spirit and body, drawn and actual: all are fused in

Grotte Cosquer. These great auks may have been food, or the source of oil and feathers, but they were more than that, alive in the mind, co-beings on earth.

Francesco d'Errico makes one further suggestion. 'All the breeding colonies recorded over the five centuries before extinction were on very small islands,' he has written.

> And all the postglacial sites which yielded remains of young specimens were on little islands or in regions richly scattered with small islands. It is probable the Mediterranean breeding sites were on similar islands, far enough from the coast to discourage swimming predators. Palaeolithic people would have reached them by some means of sea transport. If Cosquer depicts an encounter between males disputing a female on a breeding site, that must have been seen by Palaeolithic man at a nesting site during breeding time.

The implication is clear: these images are evidence of man 20,000 years ago having the ability to cross open water to distant islands. The birds shown breeding here were remote creatures, difficult of access, in that way rare and precious. The horses, bison, aurochsen and deer on the same walls were, at this very moment, in the process of being hunted out. Perhaps there was a recognition here that already, 20,000 years ago, the great auks had also begun to move down that path and that the Anthropocene, the era in which man was shaping the destiny of the earth, had already begun. And so the first depiction of seabirds was not something early in the story of man's relationship with other animals, but late, perhaps with the end already in view. The Cosquer auks are a sign that man had already hunted too well: the land-based herds had been depleted; he had been forced to devise seagoing craft with which to pursue his meat offshore; and in the charged depths of the cave itself, by torchlight and firelight, he had begun to number the great auks among the creatures he revered.

Slow, meaty, fatty and feathery: the great auk was as good a dinner as palaeolithic man could have wished for. But of all the great auk finds, the most endearing and unlikely was in the late Roman layers underneath the Plaza del Marques in Gijón in northern Spain. The bones of the bird, which must have come in from the Bay of Biscay just to the north, were surrounded by the remains of a flock of chickens. Were they simply differ-ent parts of a menu? Or did a great auk live for a while in an elegant, columned Roman coop, leading his gaggle of hens around him, clacking away at them with his giant ridged bill, king of the northern birds, treasured as a noble oddity by a provincial Roman, reading his Horace, sipping his *vinho verde*?

* * *

Traditionally, the relationship of people to seabirds was 'eat and revere'. Use them and love them. In the course of the sixteenth and seventeenth centuries, that frame of mind evolved into its modern successor: 'exploit and collect'. The first great auk to be kept as a pet lived for a while in Copenhagen before being stuffed and taking his place in a Cabinet of Curiosities, but the scale on which people came to use them in the early modern Atlantic reflected the enlarging potency of the new colonizing and commercializing drive in Europe. The capacity of the human organism to hunt and kill, evident since the first emergence of modern human beings 60,000 years ago, did not change in the seventeenth-century Atlantic but was magnified by technology and by the increasing demands of the market.

By the sixteenth century, European fishermen had already done major damage to the stocks of their own seas. Most

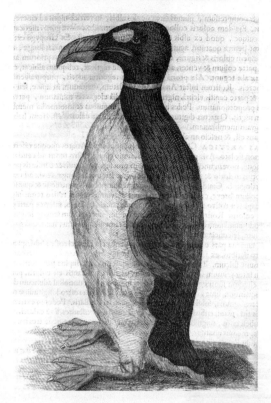

The first great auk to be kept as a pet belonged to Ole Worm, the Danish scientist. The white band around the neck was not part of its plumage but the collar that Worm used to control it on walks.

European great auks had been dead for thousands of years. At the end of a 600-year-long raid on the Atlantic, Europeans were fishing in a diminished ocean. They had forgotten what an untouched nature looked like. Fish shortages had begun to appear. In the 1580s and 1590s, the Cornish pilchard catch plummeted. Cod disappeared from the Faeroes in the 1620s and the migratory fish of European rivers had long since been decimated. The Western Atlantic beckoned both as an invitation and as a necessity.

When the first Europeans arrived off Newfoundland, the seabirds were 'as thick as stones lie in a paved street'. The Europeans, attuned to an ocean they had profoundly damaged, could not quite believe the completeness of what they encountered. Every cod, it was repeatedly said, had a coin in its mouth. This nature was money waiting to be pocketed.

Although the actual number of Europeans in America was small (about 57,000 white Americans by the 1660s), they were merely the agents of a European population that had boomed to more than 70 million over the previous hundred years. Famine remained a European reality in the seventeenth and eighteenth centuries for a continent that had depleted its own resources. The true picture, then, is not of a few pioneers arriving on American shores, but a hungry continent extending its reach across the ocean. In that sense, the whole of Europe had arrived in the virtually untouched American ecosystem, ready to consume it. By the 1570s, fifty English vessels, a hundred from Spain, another thirty from 'Biskaie' and others from Brittany and Portugal were fishing for the Newfoundland cod, anchovies, squid and whales. By the middle of the seventeenth century, as the scientists in Europe were beginning to tabulate and codify the natural world, an astonishing 200,000 tons of cod were being shipped from Newfoundland to European and Mediterranean markets every year.

The birds were implicated in this raid from the start. In 1534 Jacques Cartier, the great French explorer, was overwhelmed by the sheer abundance, particularly of great auks. Funk Island, off the Newfoundland coast, was 'so exceeding full of birds that all the ships of France might load a cargo of them without any one perceiving that any had been removed'. In less than half an hour Cartier's men filled two boats with their bodies. They and the birds from numerous islands and egg rocks between Cape Cod and Newfoundland could help with the fishing. 'The fishermen doe bait their hooks with the quarters of Sea-fowle,' the new colonists reported. Shearwaters, caught and sliced, became the favourite bait for fishermen so that with the rise of commercial fishing these island sanctuaries away from the mainland became slaughter-houses. As early as 1578, the Englishman Antony Parkhurst reported that 'The Frenchmen that fish near the grand baie doe bring small store of flesh with them, but victuall themselves always with these birds.' The bodies were barrelled up with salt, eaten instead of salt pork. The flightless auks were said to be driven on planks or sails into the boats tied alongside the rocks. It is impossible to imagine any modern puffin, guillemot or razorbill being driven in that way but flightlessness would have had a profound effect on the great auk's psyche and it is not impossible that they were herded towards their doom.

To collect their bait, sailors would draw in crowds of storm petrels.

The most common and effective way of killing them was with a whip, which was made by tying several parts of cod line (each part 7 or 8 feet long) to a staff 5 or 6 feet in length. The petrels were tolled up by throwing out a large piece of codfish liver, and when they had gathered in a dense mass, huddling over the object which attracted them, swish went the thongs of the whip, cutting their way through the crowded flock and perhaps killing or maiming a score or more at a single sweep. By the time these were picked up another flock was gathered, and the cruel work went on until, maybe, 400 or 500 birds were killed, though perhaps it was seldom that so great a number was obtained at once.

The depredations gradually had their impact. By the 1770s, the 'bird islands are so continually robbed, that the poor Indians must now find it much more difficult than before to procure provisions'. By the 1830s the colonies between Cape Cod and Newfoundland had been so deeply depleted that people in search of seabird eggs were forced to go as far north as Labrador each summer. In 1816 the Reverend Jonathan Cogswell of Freeport could say that 'Birds of no kind abound in Maine.' Local extinction had done its work. The result was that by 1800 the north-west Atlantic was beginning to resemble European seas, an environment brutally transformed by the arrival en masse of relatively few technologically equipped and ideologically empowered Europeans. The growth of the population and the urbaniza- tion of the shore had added nutrients to the system that had changed the whole micro-environment for the plankton and algae. Coastal and river habitats had been destroyed and polluted. Untold quantities of fish and birds had been taken out by the super-predators. In every way – from 'top-down' exploitation, 'bottom-up' nutrient loading and 'side-in' habi- tat destruction and pollution – the pre-existing world had been changed. 'Through these multiple, interacting impacts,' as the Canadian marine biologists Heike Lotze and Inka Milewski have written, 'humans increasingly dominated the coastal ecosystem, inducing shifts in food web composition and species interactions on every trophic level.' At its deep- est biological level, the Europeans had Europeanized the new world. It was probably no longer a place in which the great auk could survive.

By the end of the eighteenth century, the wrongness of this destruction had started to become apparent. Aaron Thomas was the thirty-two-year-old son of a Herefordshire farmer. His brothers were London merchants and silversmiths, but he was no more than an able seaman on HMS *Boston* when he arrived in St John's, Newfoundland, in 1794. Thomas was literate, may have been a purser's assistant and of his journey to Newfoundland wrote a long account in 'clear bold handwriting on good paper'.

Newfoundland had always been overwhelmingly male, tough, drunk and violent. There were so few women in the mid-eighteenth century that the English scientist Sir Joseph Banks had noticed that even his washerwoman received an official invitation to the governor's ball. And Thomas witnessed the mindless cruelty of those who had come to consider animals as things: a goose was used to clean out a chimney; another to stop a chimney fire. For fun, a live lobster was clipped to a cat, which went into a frenzy, jumped on to a horse, where it dug in its claws, and to general amusement the horse careered away in pain and fear.

When listening to fishermen's tales, Thomas heard of the usual way of treating the great auks:

> If you go to the Funks for Eggs, to be certain of getting them fresh, you pursue the following rule: – you drive, knock and Shove the poor Penguins in Heaps! You then scrape all the Eggs in Tumps in the same manner you would a Heap of Apples in an Orchard in Herefordshire. Numbers of the Eggs, from being dropped some time, are stale and useless, but you have cleared a space of ground the circumference of which is equal to the quantity of eggs you want, you retire for a day or two behind some Rock at the end of which time you will find plenty of Eggs – fresh for certain – on the place where before you had cleared.
>
> If you come for their feathers do not give yourself the trouble of killing them but lay hold of one and pluck off the best of the Feathers. Then you turn the poor penguin adrift,

with his skin half naked and torn off, to perish at his leasure. This is not a very humane method, but it is the common Practize.

While you abide on this island you are in the constant practize of horrid crueltys for you not only skin them alive, but you burn them Alive also, to cook their bodies with. You take a kettle with you into which you put a penguin or two, kindle a fire under it, and this fire is absolutely made of the unfortunate Penguins themselves. Their bodies being oily soon produce a Flame; there is no wood on the island.

Seabirds as fuel to boil their own kind; birds stripped naked to perish at their leisure; incubated eggs bundled into heaps. Aaron Thomas did not belong with what he described. As a man of sensitivity and intelligence, he was the liminal witness, observing one set of behaviours from the assumptions of another. Nevertheless, these scenes represent the pit of the materialization of the natural world, the folding out into practice of the Cartesian separation of body and spirit. And the great auk was its inevitable victim.

Modern human beings did terminal damage to the great auks, but the catastrophe may have had deeper roots. Hunters in early Europe and pre-contact America certainly reduced their range; many great auk bones have been found in ancient middens in Norway and the Atlantic seaboard of America in places from which it was absent after 1500. In Norway, the oldest bones are from 11,000 BC and the youngest about 10,000 years after that. In any rubbish tips in Norway that are later, there are no great auks. It seems likely that the Norwegians had killed off these birds by the end of the Bronze Age. What was described in the last few hundred years on the margins of the ancient territory of the great auk was almost certainly just the tail end of the process.

Because the mid-Victorian moment at which the great auk disappeared was the age of the relic, it had an extraordinary afterlife. No bird has lived more dead. The Victorians made a fetish of absence. Elegy, not celebration, took its place at the heart of their idea of beauty. Nothing was more to be treasured than the emotional charge around the ending of things, and Victorian beauty found its refuge in the dying fall. This may have been, as the American critic Deborah Lutz has suggested, 'an anxious grasping after permanence' in an age of rapid change. But there was more to it than that, a desire, as Lutz had also said, 'to keep loss palpable, not to be consoled but to feel the death – annihilation – as inspiration'.

Alfred Tennyson's *In Memoriam*, his great sequence of lost love for his friend Arthur Hallam, bridges the years in which the great auk finally died. Hallam collapsed suddenly in 1833; the last great auks ever seen for certain were killed in 1844; Tennyson finished *In Memoriam* in 1849 and the poems might be seen as an inadvertent commentary on extinction as the core of existence. Time, for Tennyson, had become 'a maniac scattering dust', the generator of meaninglessness:

So runs my dream, but what am I?
An infant crying in the night
An infant crying for the light
And with no language but a cry.

Famously, Tennyson held out some consolation for those confronted with the loss and disappearance of what had once been inexpressibly valuable:

I hold it true, whate'er befall;
I feel it when I sorrow most;
'Tis better to have loved and lost
Than never to have loved at all.

But buried in those lines is something that may have been truer still:

'Tis better to have lost and loved
Than never to have lost at all.

Loss, for these Victorians, had come to look like value itself, and nothing embodied that more powerfully than the story of the great auk and, even more poignantly, its egg. Great auk eggs became a kind of shrine of loss for collectors all over Europe, fragile, perfect, enormous, ironically beautiful as the beginning of something that would never begin again. No one paid more attention to the great auk and its egg than Alfred Newton, professor of comparative anatomy at Cambridge, and the man he loved and admired more than any other, John Wolley.

Newton had been lame since childhood, from playing too rough a game when a boy, and to the end of his life walked with a pair of sticks. Wolley had been a hero to all who knew him, a mountaineer and rock climber, vastly brave, who had kept snakes in his desk at Eton, could swim like an otter and was 'of undaunted courage and adventurousness'. Everything Newton, who was inward and quiet, might long for was to be found in his friend. The north held an attraction for them both, 'the land of the gyrfalcons and the capercaillies, of bears and wolves', as Newton described it. Together in 1858 they went to Iceland to seek, but not to find, the last signs of the great auk. A year after their fruitless return, Wolley was suddenly overcome by a terrible lassitude, perhaps a viral infection, perhaps a tumour, and he died aged thirty-six on 20 November 1859.

Newton was devastated. It was his own *In Memoriam* loss of all promise of love, happiness and fulfilment. Newton never married, and was actively misogynistic, appalled when women were first admitted to the chapel of Magdalene College, Cambridge, where he was fellow, but there is no need to sexualize this story. Wolley was by any account as lovable a human being as Arthur Hallam had been, the man, as it says on his memorial tablet in Southwell Minster, 'who brought for the first time into England the eggs of the smew, the waxwing, and the broad-billed sandpiper'. As he was dying, he made it clear that his great collection of eggs, gathered from all over northern Europe, should be given to Newton. In February 1860 the collection arrived at Elveden, Newton's father's house in Suffolk: 'twenty-four enormous packages, which weighed altogether one ton and filled a railway truck'.

Over the next fifty years, Newton worshipped at the shrine of the lost man and the lost bird. Slowly and piously he elaborated a fascinating book around Wolley's egg collection, finally published towards the end of his life as *Ootheca Wolleyana* in two stupendous volumes, meticulously detailed, beautifully illustrated with paintings of the prize eggs by Henrik Grönvold, the great Danish bird and egg painter of the age, and filled with Wolley's own wild adventures up crags and into the nests of eagles and gyrfalcons, one of the most illuminating and life-filled windows into the Victorian frame of mind.

No one in Europe collected more of the great auk eggs than Newton. No one pursued their history more assiduously around the continent, researching their origins and their fate in auction houses and drawing-rooms, in hidden cabinets and unsuspected drawers, accumulating layer on layer of information about any fragment that might subsist of egg, skin or bone. With the curator's exactness and the obsessive's repetition, feeding the unassuageable hunger to gather up the loss, Newton filled file after file with his notes in tiny crabbed writing on the extinct bird, all now carefully archived in the University Library at Cambridge to which he left them.

Those files also contain some of the most beautiful photographs of seabird eggs ever made. The great auk eggs varied in size, the biggest up to 3 inches across and more than 5 inches long, half as big again as a swan's, immensely yolky in life, somewhere between oval and pear-shaped, with the ground-colour and markings varying as they do with the guillemot and razorbill, each mother bird laying the same pattern year after year. Newton's photographs of them are taken at eye level, the eggs seen as perfect and monumental, temples of loss: Mrs Henry Wise's egg, bought by her father of Charlton Court, Steyning, spotted all over, sold in 1888 for £225; John Wolley's egg, scribbled and blotched; Newton's own egg, very bare; Lord Lilford's eggs, now in Cambridge, fully and regularly patterned all over, the most beautiful; the Pimperne eggs of Mrs Hill, with large blotches on the shell like the continents and seas on the moon; the eggs Newton had discovered lying unsuspected in the filing drawers of the Royal College of Surgeons now identified for what they were covered in fine Twombly-like scribbling.

The melancholy of loss hangs over every photograph. On the flyleaf of the notebook that John Wolley had kept in Iceland, when he and Newton had been hoping to find the last living examples of the great auk, Newton later inscribed lines from *Lycidas*, Milton's great elegy on his young friend who had died at sea: 'For Lycidas is dead, dead ere his prime, / Young Lycidas, and hath not left his peer.' The world of the great auk was now the non-future. Each of these eggs has its own topography or biography, covered in delicate and impenetrable hieroglyphics, some patterned as if made of scagliola, others visibly porous, or planetary, or as if burnt, their focus blurred in the ancient photography. It is easy enough to be drawn into their world. Some look like the night sky in negative, with dark meteor trails across their black-on-white star maps. The Walter Rothschild egg is smooth and plutocratic, perfect but featureless, the Earl of Derby's dark and dense with jots and doodles.

These eggs were as dispersed across the world as the birds had once been across the Atlantic. They went on their own odysseys. One famous egg had the word '*Pingouin*' painted carefully on its shell, said to have been written there by a Monsieur Dufresne, keeper of the Cabinet of Natural History belonging to the Empress Josephine. It somehow travelled to Leiden and from there migrated to a famous collector, Monsieur Josse Hardy of Dieppe, a shipowner and birdfancier. Hardy left it to his son Michel who loaned it to the Dieppe museum. Michel's daughter, Madame Ussel of Eu, put the egg up for auction at Stevens's Auction Rooms of Covent Garden in London in 1909. 'With the financial support of Lord Strathcona', Mr Hay Fenton of Lombard Street, London bought the egg for 190 guineas and two days later presented it to the Natural History Department of Aberdeen University, where it still is.

The prices climbed steadily through the nineteenth century, a sombre re-run of seventeenth-century tulipomania: £15 15s 6d for an egg in 1832, a price nearly double the average annual income for a skilled worker; £30 for an egg in 1865, £64 in 1876, £107 in 1880, £168 in 1887, £225 in 1888 and topping out at £315 in 1894, when the average annual income was £83. Eggs were displayed at restaurant dinners. Frauds were practised. Newton's collection at Cambridge now contains a large tropical bean which someone had tried to pass off as the egg of a great auk. Newton himself was marginally involved in a shady deal by which a repainted swan's egg was substituted in a collection for a particularly handsome great auk's egg. At the end of the century, the Edinburgh Castle, a pub in Mornington Road, Regent's Park, boasted 'no fewer than three eggs of the Great Auk, whose aggregate cost at auction was 620 guineas'. (Those eggs are now at Harvard.) The Queen owned a jewelled great auk egg made by Fabergé. Great auk cigarettes would be smoked in the trenches. The Sèvres porcelain factory produced a great auk dinner service. The footman of Lord Garvagh, an Irish peer, enjoyed passing fame as the man who had dropped and smashed his employer's great auk egg; he was dismissed. Prague, Copenhagen, Amiens, Abbeville, Autun, Paris, Rouen, Strasbourg, Berlin, Dresden, Hanover, Dublin, Florence, Milan, Pisa, Turin, St Petersburg, Oslo, Stockholm, Lausanne, Chicago, New York, Los Angeles and Cincinnati: all these and more had their great auks or their eggs with which to make real to their owners or citizens the poignancy of extinction.

They were fetishizing absence. Newton's life outside the confines of memory and the gatherings of the great bird was desiccated. His lectures were 'desperately dry and very formal'. He read them out, sitting down, from a word-for-word script that was marked at regular intervals by a small drawing of a glass. Here, Newton would pause and take a sip of water before continuing. Even when only one student was present, Newton would stick to what he had written down. 'Most of my audience are well aware ...'

Newton was planning, for twenty-five years or so, the great book on the great auk. To his horror, in 1885, an amateur naturalist called Symington Grieve, who had been gathering information in his spare time, beat him to it and published *The Great Auk, or Garefowl*, inevitably full of what Newton considered the inability 'to distinguish between good and bad'. Newton never wrote his own.

His memorial is now the collection of great auk eggs gathered deep in the neon-lit basement of the Cambridge University Zoology Museum. Each egg is a mausoleum, a little cromlech to forgotten lives, settled in its curatorial nest inside a grey modern filing cabinet. It is all a million miles from the stink and rage of any bird colony or the roar of the Atlantic. These things gathered from the past are mostly smooth to touch, stony, marble almost, unbelievably precious in the hand, sacramental. The shell, where you can see it through the holes drilled in the ends where the eggs were blown, is one twenty-fifth of an inch or so thick, robust, not fragile, as monumental as the idea of the great auk would require it to be.

The patterns are entrancing; you can feel yourself being pulled into their allure: some of the scrawling and doodling is the colour of dried blood, some greenish, others drawn all over as if with a wayward pen in zigzags and loops. Some of the eggs are rounder than others, some adamantine smooth. The spots can look as if they have been dabbed on in blue-black ink on the grainy cream ground. Others are fading into a bland lunar simplicity.

*   *   *

These eggs were laid and the chicks raised on shallow rocks on to which the auk could haul itself out like a seal or a penguin before standing upright and walking perhaps a short distance inland. For a week or two, the chicks would have been fed on land. A guillemot chick needs to be warmed by a parent for about ten days, but after that its own circulation will keep it warm, and so ten days may be about the minimum a great auk chick would have stayed ashore. Penguins regurgitate food from their stomach for their chicks. It is one way a large catch can be carried ashore and although there is some loss in goodness, because the parent has begun to digest the fish, there is no other way to get the food to the chick. It may be that the great auk fed its chick in the same way. Not for long, though. Like the razorbill, it almost certainly took its chicks out to sea when two or three weeks old, as it would have been unable to travel far to find fish and the economics of an early move to the sea would have made sense. A fully grown great auk may have needed 2 pounds of fish a day for itself, apart from anything needed for the chick, and so that is a scene to imagine when at the razorbill colony today: the great auk fathers growling and croaking to their giant anxious chicks, urging and inviting them away into the Atlantic for the life to come.

Great auks specialized as deep-diving fish-feeders, fast in the water. They could outpace a six-oared boat when on the surface, swimming 'with their heads much lifted up, but their necks drawn in'. Like the razorbills, they dived when alarmed. It might be tempting to think that flightlessness was the great auk's handicap but flightlessness in seabirds has often been a powerful choice. Members of five different lineages – the cormorants, auks, penguins and ducks, plus the big northern hemisphere cormorant-like divers called Hesperornithiformes, some of which were 6 feet long – have all chosen the flightless life. The Hesperornithiformes did not survive the mass extinction 60 million years ago, but many other large flightless birds, including at least eight different species of flightless puffins in California, swam in the ocean until about 20 million years ago, when some even more efficient competitors came into the sea to take their place: the seals and sea lions.

Flightlessness can be a good life choice for one simple reason: bodies that are good for flying are bad for diving, and

vice versa. A wing that is designed for the best possible performance when under water – one that is on its way to being a fin
– leads to enormous expenditures of energy when flying. If
you are a diving bird, flight will allow you to escape, but the
trade-off is that you won't be as good a diver as you would
otherwise have been. If you can be safe, then flightlessness is
probably going to be an advantage, even if you still have to
negotiate the seals, whales and sharks that would like to eat
you. Penguins, for example, walk and swim vast distances
between breeding colony and food sources, journeys that
would be much more easily made by flying. The benefits of
flightlessness counteract that cost.

Great auks were no relation of the penguins, which had
diverged from them at least 80 million years before, but their
bodies had to come to resemble each other because environmental logic had pushed them together. Penguins have reduced
their wingspan, enlarged and strengthened their wing bones,
increased the weight of their bodies, developed muscles which
work best at low, underwater wingbeat frequencies, and
increased the myoglobin, oxygen-retaining content of their
muscles so that they can hold their breath for longer under
water.

If you want to fly, all the logic pushes in the other direction:
you need longer wings, lighter bones, a lower overall weight
and faster-working muscles. Try clapping your hands
underwater and you will see that what is suitable for one
medium is no good for the other. Most seabirds make the
compromise, but as Kyle Elliott from McGill University has
shown, there comes a point where compromise is no longer
possible. The big northern Brünnich's guillemot or thick-
billed murre is right at the limit of a flying diver. Its flight costs
– the energy needed to keep airborne – and its dive costs – the
rate of oxygen consumption during dives – are both nearly

unworkable. These guillemots have to keep their wing size down to make them effective under water with the result that they have the highest ratio of body-weight to wing area of any flying bird. Because of that tiny wing, flying for the guillemot is intensely exhausting, their metabolism in flight running at an astonishing thirty-one times the rate it runs when they are sitting calmly on the ledge with their chick. No other vertebrate performing an intense activity has been found with a multiple of more than twenty-five.

If life is safe enough – on an offshore island with no predators, or surrounded by many similar birds that would distract a predator – then flightlessness is an obvious option. Bodies that don't have to fly can grow larger and swim deeper for longer. They can insulate themselves with fat rather than air, reducing their buoyancy and making diving easier. They can live, in effect, like seals. That is the choice the great auk made about 20 million years ago when it diverged from the razorbills and started on the road to magnificence, enlarging quickly, fattening, triumphing, not as the sad, stupid, vulnerable cul-de-sac of the evolutionary process, but as a masterpiece of beauty and adaptation, as heroic as the gyrfalcons, regal in its presence across the northern seas.

Penguins can dive to 1,850 feet and stay under for twenty minutes on a single breath. That is how to imagine the great auk, deep in among the riches of the ocean fronts, feasting on the meeting of the Labrador Current and the Gulf Stream, the emperor of its dark cold prairie. Only when the human super-predator arrived with his ocean-straddling technology and a voracious European market at his heels did the ancient genetic choice of flightlessness become a liability, and in about 1844 one more bird species joined the 140,000 that had already succumbed to extinction.

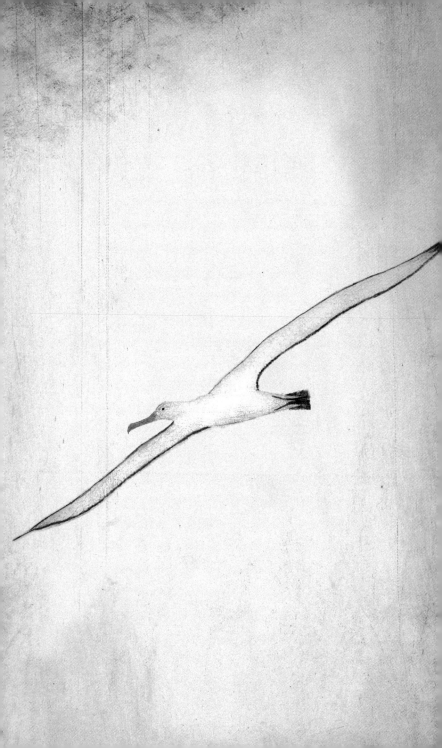

10

Albatross

======

No one has ever encountered the full burning ecstatic beauty of a seabird quite in the way the twenty-two-year-old Herman Melville, crewing as a green hand on board a New Bedford whaler deep in the South Pacific at some time in 1841, first met an albatross.

It was during a prolonged gale, in waters hard upon the Antarctic seas. From my forenoon watch below, I ascended to the overclouded deck; and there, dashed upon the main hatches, I saw a regal, feathery thing of unspotted whiteness, and with a hooked, Roman bill sublime. At intervals, it arched forth its vast archangel wings, as if to embrace some holy ark. Wondrous flutterings and throbbings shook it. Though bodily unharmed, it uttered cries, as some king's ghost in supernatural distress. Through its inexpressible, strange eyes, methought I peeped to secrets which took hold of God. As Abraham before the angels, I bowed myself; the white thing was so white, its wings so wide, and in those for ever exiled waters, I had lost the miserable warping memories of traditions and of towns. Long I gazed at that prodigy of plumage.

The creature that had arrived in front of Melville as if from another universe, stripped of any associations with any life he knew, must have been the biggest of them all, a wandering albatross, whose wingspan is 11 feet or more, and of which the males begin their lives pied and spotted, but which, as they age, grow paler with every passing decade until they die as pure, unearthly and archangel-white as the bird Melville saw, maybe seventy or eighty years old, having fathered forty or fifty offspring and flown millions of miles across the Southern Ocean.

A young wandering albatross, yet to develop the pure whiteness
Melville saw

That otherworldliness is what we all see in them now. I have stood and watched for hours at the stern of a supply ship as it made its way across the Southern Ocean from South Georgia to Port Stanley in the Falklands. Wrapped in every piece of clothing I had, I always chose the lowest deck so that now and then, as the stern dropped into a trough, the bright, lit water came aboard, across the red-painted steel of the deck and over my boots so that I had to hang for a moment from the stanchions until we rose again on the swell and the wide, windstrewn landscape of that ocean reappeared, a storm-world stretched from one horizon to the other. Beside, behind and above us, along with the beautiful chequered cape petrels, domino birds, patterned like flying fragments of a draughtsboard, the albatrosses hung and floated for hours and days. They sailed beside us without effort, gliding across the swells as if all directions were equally possible, alive on an ocean of potentiality, alert on their seamless ice-surface of ease and command.

Only rarely, and more often in the distance than not, a wandering albatross made its way across the ocean, appearing and disappearing as if we were merely an incident in the vast compass of its life. Seen in one moment, it would be gone in the next. The commonest was the black-browed albatross, the mollymawk (from the old Dutch word for midges), with dark smudged mascara lines around its eyes; the most beautiful the light-mantled sooty albatross, given the name *pio* or *piew* by the old sailors, after the dropping tragic note of its call.

These sooty albatrosses like to follow ships, maybe because evolution has encouraged them to follow whales across the ocean, scavenging the remains of squid or fish that fall from the mouth of those messy eaters. And when the wind is not with them they can use it all the same to pursue their course, to lift them 50 or 60 feet above the sea, and then allowing gravity to glide them downwards in the direction they need to go. This dynamic soaring is a constant and beautiful rhythmic alternation, the slow metronomic pulse of an ocean voyager's life, wind power up and gravity down, zigzagging and tacking in the wake of the ship, or up alongside, eye level to eye level with any human, mesmerized watcher. Four or five times a minute, you can see a bird cruising down in the trough between swells and then emerging on to the lip of a long roller suddenly finding itself blown up and backwards in the gale from which the wave had sheltered it, using that surge of wind energy to lift it to the point where it can continue on its chosen path. Physics, geometry, acrobatics, insouciance, liquidity, fluency and calm: the constituents of archangel genius.

Amazement at their command has been the one constant in any human interactions with the albatrosses. Peter Mundy, a London East India merchant, met the enormous birds in the South Atlantic in June 1655:

> It maketh mee wonder how these and others [fly] with continuall steaddy outtstretched wings, nott seeming to us to moove them att all in a long space, as wee see hawkes and kites to soare aloft: some allead[g]ing the depth of aire supports them, butt these flew close to the water and never farre from it.

In July 1912, the great early twentieth-century American seabird ornithologist Robert Cushman Murphy, fresh out of Brown University, had been instructed by the American Museum of Natural History to go on the *Daisy*, a New Bedford whaling brig, as assistant navigator. They were to sail down into the high latitudes of the South Atlantic hunting for sperm whales and to South Georgia for the oil of elephant seals. Murphy, like all sailors before and after him, watched entranced.

> On November 18 [1912] in Latitude 48° 49′ south and 36° 40′ west [about 200 miles north of South Georgia], we lay to all day, in the teeth of a southwesterly gale ... During the morning, four Sooty Albatrosses joined us and remained near-by for five hours, appearing to have no other purpose

than to play in the howling wind for the admiration of us on board. Their ease and precision, and particularly their ability to vary their speed and to 'stand still' in the air, put them in a class by themselves. To scratch their polls with the claws caused no hitch in their gyrations, and one of them was seen to turn down its head and preen the feathers of its belly without losing its place in the squad.

The four sooty albatrosses stayed with them for day after day. These sombre and beautiful birds, their pale, pearly bodies contrasted with a dark cap and wings, their eyes startlingly outlined in a white ring, now and then making what Murphy called their 'hollow ghostly trumpeting', are probably the most famous albatrosses that have ever lived. The light-mantled sooty albatross may well be the bird that lies behind Coleridge's great poem of the mariner and the albatross. But in nearly all depictions of that poem, the artists assume that Coleridge had in mind the giant-winged wandering albatross. It is usually made to hang from his neck with a body almost as large as his own, the weight of its vastness the intolerable consequence of a crime against nature.

That may have been the direction in which Coleridge pushed the story but it is not how it began. Although he later denied it, he took the idea of the sailor killing the albatross, thereby setting in train the poem's sequence of guilt, suffering and redemption, from a suggestion made to him by Wordsworth, when out walking together in Somerset in the autumn of 1797. Wordsworth had been reading the account of a troubled and violent round-the-world voyage made between 1719 and 1722 by an English mariner, George Shelvocke. In that narrative, Shelvocke tells how in late September 1719 his ship, the *Speedwell*, was driven far south of Cape Horn, deep into the world of ice and fogs. When the men of the *Speedwell* managed to catch sight of land at all, they were given a 'melancholy prospect of the most desolate country (to all appearance) that can be conceiv'd'. Fierce tidal streams gripped their hull, driving them backwards even as they were sailing at 6 knots into it, past 'a most uncomfortable landskip ... cover'd with snow to the very wash of the sea, [which] bears more of a likeness of a huge white cloud, than of firm land'. Their masts were cased in ice as they were taken down to 61° South, in prodigious seas and continual misty weather 'which laid us under hourly apprehensions of falling foul of Islands of ice'. This was the world in which the light-mantled sooty albatross was at home.

The English sailors tried to make everything 'as snug as possible', but one man, on 1 October, found his hands and fingers so numbed he could not hold himself while furling the mainsail and 'before those that were next to him could come to his assistance he fell down and drown'd'.

One would think it impossible that any living thing could subsist in so [f]rigid a climate; and, indeed, we all observed, that we had not the sight of one fish of any kind, since we were come to the Southwards ... nor one sea-bird, except a disconsolate black *Albitross*, who accompanied us for several days, hovering about us as if he had lost himself, till Hatley, (my second Captain) observing, in one of his melancholy fits, that this bird was always hovering near us, imagin'd, from his colour, that it might be some ill omen. That which, I suppose, induced him the more to encourage his superstition, was the continued series of contrary tempestuous winds, which had oppres'd us ever since we had got into this sea. But be that as it would, he, after some fruitless attempts, at length, shot the Albitross, not doubting (perhaps) that we should have a fair wind after it.

That is the germ of the most famous seabird story in the world: not the great white albatross, but the smaller beautiful dark one, no more than 6 or 7 feet across its wings; shot because its presence seemed like a bad omen; for a crew disturbed by endless fog and ice, and by the threat of icebergs appearing out of banks of mist that could look like icebergs themselves.

The light-mantled sooty albatross, perhaps the bird at the heart of *The Ancient Mariner*

Coleridge changed everything that Wordsworth and Shevlocke had given him. The Ancient Mariner's albatross is as white as Melville's. Its death, not its life, brings bad luck. And its shooting – with a crossbow, not a musket, as Coleridge wanted to plunge his story deep into the medieval past – is made to seem not like the excision of a bad spirit from the threatening ice- and fog-bound skies of the Southern Ocean but a crime against nature, a transgression against the order of goodness in the world. That was Coleridge's great and astonishing invention. The act, he wrote, was 'in contempt of the laws of hospitality'. No one before him had thought it wrong to kill an albatross or any seabird. His flipping of black to white and of victimhood to guilt was, at a stroke, the invention of the modern ecological sensibility. For the first time, in *The Ancient Mariner* the moral centre of virtue came to lie not in the thinking and acting human being but in the world to which he does damage. It is an early intimation of Jakob von Uexküll's recognition a century later that the human *Umwelt* – our way of encountering and perceiving the world – is not the only *Umwelt* there is.

That revolutionary change was a thousand miles away from anything Coleridge's contemporaries might have thought. Captain Cook on his three great voyages to the South Seas between 1768 and 1780 had often seen 'Many Albatrosses about the Ship, some of which we caught with Hook and line'. The birds 'were not thought dispiseable food even at a time when all hands were served fresh Mutton'. The scientist Joseph Banks, on board for the first voyage, was happy on 5 February 1769 to eat 'part of the Albatrosses shot on the third, which were so good that every body commended and Eat heartily of them tho there was fresh pork upon the table'.

Nothing superstitious hung about the killing of a pre-Coleridgean albatross. In fact, the killing of albatrosses was standard fare for all voyagers into the South Seas through the nineteenth century.

They were shot or caught by a baited hook on a long line. 'Its webbed feet make capital tobacco-pouches … The wing-bones make excellent pipe-stems; the breast … a warm though somewhat conspicuous muff; and the beak, in the hands of a skilled artificer, a handsome paper-clip.' The wandering albatross was known as a Cape Sheep among sailors heading around the South Africa: its feathered skin was useful as a kind of fleece.

'Warm reception that ye first albatross met with', a drawing by Edward
Moore, emigrating to Australia, 1854

Even the great Robert Cushman Murphy, a man of immense
humanity, intelligence and grace, was the heir to that tradition
when collecting for the American Museum of Natural History
on the *Daisy*. He loved the sooty albatrosses as much as any
man ever has and for days after the storm in which the four
birds had attached themselves to the brig, he watched them as
if they were his own.

'The quartet was with us mostly during late afternoon each
day,' he wrote years later. But he was at sea to gather material
for the museum.

On the evening of November 20 [1912], I shot one of these
Sooties as the four of them crossed our deck amidships. The
victim tumbled into the arms of a sailor, nearly knocking
him down. Shortly after noon next day, in latitude 51° S., a
group of *three* Sooty Albatrosses overtook us and drilled in
the usual formation. The same was again true on November
22; and in late afternoon of the 23rd, just after we had made
out the white highlands of South Georgia, the familiar trio

again appeared, encircling the brig and poising successively
on the tip of the bowsprit and over the gilded truck of the
fore-topgallant mast.

The flesh, Murphy thought, was 'sweet not firm'; the egg 'like
beavertail soup'.

Whalers, filling many dull whaleless days at sea, often
caught them on hook and line. The hook could not penetrate
the hard plates of the bill and the trick was to keep the line
taut as you pulled the bird in. Cleverly done, the bird could be
brought on deck with no harm. Sometimes, wanting to know
if the birds that seemed to be accompanying them were in fact
the same individuals, they painted them with tar before releasing them, and so establishing in the loneliness and melancholy
of the voyage some kind of warmth and connection. Others
tried to keep albatrosses as pets. The frequent sightings of
southern hemisphere albatrosses in the North Atlantic –
including at Spitsbergen, in the Faeroes, in Antwerp and
Dieppe – may have been these pet birds brought to the wrong
hemisphere and released late in the voyage. Others used them
as postmen. Melville's captain, after the young novelist had
drunk his fill of 'the mystic thing', did just that, 'tying a
lettered, leathern tally round its neck, with the ship's time and
place; and then letting it escape'. And the museum at Brown
University preserves 'a manuscript taken from a vial found
tied to the neck of a wandering albatross shot off coast of Chile
on December 20 1847'. The message is itself a tiny capsule of
life on a desperate whale-ship, deep in the bone-chilling South
Pacific, halfway between New Zealand and Chile:

Dec 8th, 1847. Ship 'Euphrates', Edwards, 16 months out,
2300 barrels of oil, 150 of it sperm. I have not seen a whale for
4 months. Lat. 43° S., long. 148° 40′ W. Thick fog, with rain.

As Murphy calculated, the bird had come 3,150 nautical miles as the crow flies in twelve days between the writing of the manuscript and the day it was shot.

For shipwrecked sailors, as the Wollongong scholar Graham Barwell has shown, these telegraphic qualities of the living bird could become more than diversions for the bored. French sailors wrecked on the utterly remote Îles Crozet in the southern Indian Ocean in 1887 attached a message to the neck of an albatross and released it in the hope that they might be rescued. Both bird and message turned up in Western Australia, where its arrival was reported in the papers. By the time a rescue mission was launched, the men had died trying to sail from one island of the archipelago to another.

The bird in flight; the bird disabled – these are the polarities around which the inherited meanings of the albatross revolved: all spirit, all flesh, all beauty, all pity, a double vision that reached its most troubling expression in *L'Albatros*, Baudelaire's masterpiece. The fastidious young man, 'dressed as neatly as a secretary at the British Embassy', had been sent on his way to India by his mother and worried stepfather, anxious to save him from the Hashish Club and the other delights that Paris might offer. After a wild gale off the Cape of Good Hope, Baudelaire met his own bird-on-deck at almost exactly the same moment Melville had met his on the far side of the world. The twenty-two-year-old American, a whaleboat crewman looking for adventure, saw the enormous throbbing beauty of it; the twenty-year-old Frenchman, exiled to India, which he never reached, jumping ship in Réunion, saw only its sadness, its indignity and the cruel indifference of the sailors around him.

Rarely has the easy fluidity of albatrosses at sea, immense and lazy, sinuous in their mastery, been more musically expressed than in Baudelaire's description of these

vastes oiseaux des mers,
Qui suivent, indolents compagnons de voyage,
Le navire glissant sur les gouffres amers.

vast birds of the seas,
Indolent companions on the voyage, who follow
The ship slipsliding over the bitter deeps.

That grace goes when the sailors catch and humiliate the bird, as it lies on deck clumsy and ashamed, with its unworldly white wings lying like useless oars beside it. Once so beautiful, now comic and ugly, once king of the azure-blue and prince of the clouds, now, like the poet, exiled and grounded, hooted at, disabled by the very wings on which he had once lived in the storms.

Le Poète est semblable au prince des nuées
Qui hante la tempête et se rit de l'archer;
Exilé sur le sol au milieu des huées,
Ses ailes de géant l'empêchent de marcher.

The Poet is like the prince of clouds
Who haunts the storm and mocks the archer;
Exiled on the ground among his taunters,
His giant wings mean he cannot stir.

How do they live when they are living in the wind? That is the core albatross question: for Coleridge, in the life of bliss before the crossbow ended it; for Melville, as the albatross flew off with its leathern tally, 'meant for man' but, as Melville thought, 'taken off in Heaven'; for Baudelaire, as the albatross slid across his principality of clouds; and for the modern seabird scientists, who have done more than anyone to answer it in the last twenty-five years.

Remote tracking of seabirds has a long history. The nineteenth-century whalers had tarred and labelled their albatrosses. From the beginning of the twentieth century, rings started to be recovered from around the shores of the Southern Ocean and informed guesses could be made about the transoceanic life of the birds. Radar in the Second World War began to detect migrating birds as ghostly images on blurry screens. In the late 1950s, miniature radio transmitters were developed, which could be attached to landbirds and tracked through hand-held or land-based receivers. The first proposal to track anything by satellite – an application for a grant to NASA – was made in 1962 by Dwain Warner, an ornithologist from the University of Minnesota, and his friend William Cochran, an electrical engineer with a genius for transistor-circuit design. Spurred on by the example of the Russians, who had put dogs into space and monitored their health in the 'dog Sputniks' by radio, Warner and Cochran had both been closely involved in radio tracking. But the scale of albatross life, the speed at which they travelled and the fact that they were overwhelmingly oceanic all meant that their lives could not be readily followed by any radio-tracking technologies.

Etienne Benson, a historian of environmental science at the University of Pennsylvania, has described how the two friends 'came up with the idea of animal tracking by satellite over a few beers during the biotelemetry conference held at the American Museum of Natural History in the spring of 1962'. In 'Space Tracks', an article published the following year in the museum's magazine *Natural History*, Warner asked: 'How can we find out where the penguins of the Antarctic go after the mating season; the routes of the wandering albatross or the Caribbean turtle; the forces governing caribou movements; the track of the Canada goose?' His answer: 'artificial satellites'.

Decades passed, with many hiccups, before that vision became a reality. The designers and funders of the satellites were more interested in monitoring the giant energy and temperature flows of ocean and atmosphere than the eccentric tracks of individual creatures. One 1960s proposal for a transponder, designed to meet NASA's painfully demanding technical specifications, was going to weigh 20 pounds, as much as an albatross, and cost $65,000, six times the annual salary of a secondary school teacher. An elk, maybe, could carry such a thing. No bird could.

In June 1979, Argos, a joint Franco-American system designed to accommodate wildlife tracking, went live on the back of two NOAA weather satellites. But still the equipment was not light enough for birds. The engineers' hi-tech specifications did not make life easy for wildlife biologists who required rough, tough, cheap and durable equipment. Progress was slow. By 1983, a transponder was developed with a miniature solar panel weighing 6 ounces, which was promising, but tests throughout the 1980s on a variety of wild birds all failed.

Two French researchers, Henri Weimerskirch and Pierre Jouventin, had been working in the southern Indian Ocean for years. Weimerskirch had first gone down to the wildly remote and brutally exposed sub-Antarctic islands as part of his national service. For five years between 1978 and 1983 he had studied the albatross species of the Îles Crozet, Kerguelen, Amsterdam and Saint-Paul – tiny remaining fragments of the French Empire, the Terres Australes Françaises – for his PhD at Montpellier. Pierre Jouventin was a veteran ringer and champion of what he described as these 'sanctuaries surrounded by the rich natural immensities' of the Southern Ocean. In Crozet, 25 million birds come to breed; in Kerguelen 400,000 tons of elephant seal haul out on the beaches. The penguins there catch twice the number of fish each year as the entire French fishing fleet. Together, Jouventin says, the islands represent 'the richest community of seabirds in the world'.

Both men realized, even in the 1970s, that they were witnessing a disaster. Ringing returns made since 1966 showed that the albatrosses were in steep decline, with adult birds dying far earlier than expected. They were responding by breeding younger than before. In the 1960s, many albatrosses did not start to breed until they were eighteen, nineteen or even twenty years old. By the late 1970s most were breeding before they were ten. But even that change was failing to halt the decline. It was particularly fierce among females. The scientists didn't know why but they guessed: the birds were almost certainly 'dying in fishing tackle and from deliberate trapping or shooting by fishermen', and more females were dying because they were fishing more often in sub-tropical waters where the fishermen were killing them.

In 1984 and 1985 in an effort to understand where the birds were fishing – and inadvertently following the practice of the nineteenth-century whalers who had tarred the birds – Weimerskirch and his team marked 3,989 albatrosses by spraying them on the breast with a portable agriculture sprayer, two colours, a red that turned pink and a yellow that went orange-brown. For about five months, until they moulted, their sprayed birds dispersed across the ocean. Trawler skippers and the crews of research vessels were asked to look out for them and report their locations. Just over 400 sightings were made, many by the trawlers, fishing on the relatively shallow waters not far from Kerguelen. The black-browed albatrosses liked to follow the boats, itself a worrying result. Most of the birds disappeared into the vastness of the Southern Ocean. One or two were spotted between Kerguelen and the Antarctic continent. Half a dozen were seen, after the breeding season was over, off Kangaroo Island in South Australia, almost 4,000 miles to the east. For the grey-headed albatrosses which they had sprayed, the ratio of birds marked to birds later sighted was 174:1. For all Weimerskirch's efforts, this was clearly not a means of understanding the albatross's life at sea.

In 1988, in the newsletter of the Argos satellite system, Weimerskirch, a warm, passionate, disciplined and intensely committed man, read about some miniaturized trackers made in Japan which had been successfully deployed on dolphins. Perhaps this was the answer. The Toyocom transmitters weighed just over 6 ounces and were about 7 inches long. They could be fitted on the birds using a ½-inch-wide harness 'made of fine goat leather with elastic sections' weighing less than an ounce. When it was on, the harness was completely covered by feathers.

In January 1989, Weimerskirch and Jouventin caught six male wandering albatrosses (males are bigger and so the weight of the tags would seem less). The birds had been guarding their nests on Possession Island in the Îles Crozet while their female partners were out fishing. One reason albatrosses are such ideal candidates for satellite tagging is that they are easy to catch and easy to catch again. Weimerskirch and Jouventin fitted the Toyocom tags, waited for the females to return and then watched as their satellite-connected males headed off into the ocean. The paper they published in *Nature* the following year buzzes with the excitement at what they had managed to pull off. It was the first time any bird had been satellite-tracked. One of their albatrosses had decided not to breed and had left the island, not returning before his batteries ran out. The other five came back and bred, apparently no worse for their burden and quietly tolerating the scientists while their harnesses were removed.

Each albatross had sent on average twelve locations a day to the Argos receivers on the NOAA satellites above them. The data had been transmitted from there to the central Argos processing station outside Toulouse and from there had been sent to the two biologists waiting anxiously in the Îles Crozet.

Oceanic scale came flooding out of the results. While their wives were incubating the eggs, the Weimerskirch albatrosses were flying up to 10,000 miles on a single foraging trip, looping out from Crozet as far as Antarctica, or past Kerguelen and on to Heard Island far to the south-east. Only in the daytime did they travel any distance, and would sit on the sea surface during the dark except when the moon was full: in a vision of life at sea which Coleridge would have loved, the albatrosses happily flew all night across a moonlit ocean. The distances covered and the speeds recorded were higher than for any estimates ever made for any albatross or any other bird. Since their positions were recorded only every two hours, these figures were certain to be underestimates.

It was already clear, in these first and revolutionary observations, that the albatrosses lived according to the wind. Consistently, they would fly downwind on the outward leg, tacking against it to get home. If by chance they were caught in the centre of a windless high, the satellites showed that they sat, stuck, floating for days almost in the same spot, waiting for the gales to return on which they could again become airborne. Only in the Southern Ocean was the meteorology strong enough, stormy enough and windy enough for the great albatrosses to have evolved. Without wind they would die.

Weimerskirch had opened the gates and others followed. In the southern spring of 1994, a mature female albatross was captured and tagged off Bellambi in New South Wales. She was a *Diomedea exulans gibsoni*, Gibson's wandering albatross, a speckle-winged sub-species named after the pioneering and much loved ornithologist and steelworker Doug Gibson, who in the 1950s had become fascinated by the Australian albatrosses and had begun to ring them. He had died aged fifty-eight in 1984 and partly in memory of him the scientists called this bird 'Mrs Gibson'.

A Gibson's albatross in the Tasman Sea

She was as big as any seabird in the world: more than 10 feet in wingspan, weighing a stone and a half, yet to breed but already something like eight or nine years old. The transmitter terminal, plus harness, case and battery, was attached to her back. The whole kit was 3 inches long, 2½ inches wide and an inch deep, and weighed in all about 3 ounces, or 1 per cent of Mrs Gibson's own weight. The transmitter was designed to communicate her position, the pressure and temperature of the air around her, plus her own pitching and rolling as she made her way across the ocean. All of this was to be sent to the NOAA satellites. At the same time, raw weather data, including, crucially, the wind speed and direction in each of Mrs Gibson's locations, was to be downloaded from those same satellites.

On 8 September 1994, primed for this space flight, Mrs Gibson set off for her journey across the Tasman Sea and the South Pacific. One question in the scientists' minds: how would she use the wind?

A series of lows was coming through. In the southern hemisphere, as warm tropical air drops south towards the pole, it spins clockwise and, as it does so, the system moves eastwards. The leading edge will carry winds coming from the north. If the centre of the low remains to the south, the northerlies are then followed by westerlies and finally southerlies. Mrs Gibson knew all this and she also knew that she could rely on a train of equivalent lows following each other across the Tasman Sea to New Zealand: one vortex after another in which the sequence of northerlies, westerlies and southerlies was as good as reliable, a windscape as intelligible as a rail network, once you had the timetable and the map, a menu of winds among which the albatross, over a long and wise life, could choose her path.

It may be that Mrs Gibson also had the ability, in her nostrils, to detect changes in barometric pressure. She knew, more than any of us know, where she was in relation to the weather systems that stretched across the ocean around her. Combined with the acute sense of smell of all tubenoses, she would have felt the low coming through, detecting the drop in pressure towards its centre and picking her course accordingly.

Only these skills can explain the month-long path which the satellite trackers revealed in September 1994. She set off in the northerlies and with them behind her immediately flew 400 miles south, to a fishing ground where, as she fished, she waited for the wind to come round to the west, as it always does and as she knew it would. Once it had shifted, she used it, not as a tail wind but on a broad reach, with the big storm-force winds, blowing a steady 50 knots, coming over her left shoulder, to fly back north on a curving track that took her to within 100 miles of where she had begun.

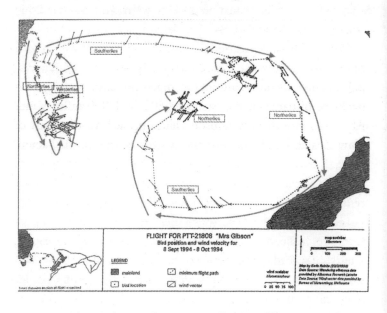

Mrs Gibson makes use of a succession of lows to sail her way around the Tasman Sea. (Reproduced from Reinke *et al.* (1998), with permission from CSIRO Publishing.)

It was a neat loop, out and back on the first two limbs of a depression. But the deep low had now moved eastwards and strong southerlies were blowing across Mrs Gibson's path. She took immediate advantage of them, keeping them just behind her right shoulder and driving hard and fast more than 800 miles to the east, ending up to the north-west of New Zealand. By then that first low had run its course and the first sign of its successor made itself apparent in winds again from the north-west. Using them, Mrs Gibson repeated her pattern: down towards the coast of New Zealand, pausing there momentarily as the low swung through and then picking up the southerlies, keeping them this time over her left shoulder, out into the Tasman Sea and then cutting north on the tail of that depression before again sitting out the beginning of the next big storm, as gale-force winds again filled in from the north.

Who might have guessed that the birds you see from the rail of your ship would live so commandingly with the wind, gyrating through all the evolutions it made possible, turning in a necklace of dance-steps, each one 500 or 1,000 miles wide? After this month, Mrs Gibson went on, back to the coast of New Zealand and out into the widths of the South Pacific, further east of New Zealand than New Zealand is from Australia, spinning around the circus of winds which the long sequence of ocean lows was repeatedly building for her.

Of all the intersections of poetry and science, here is one of the strangest. Mrs Gibson was chased south by the north winds, first as she left the Australian coast and for the second time off New Zealand; and then, turning, twice, she made her northing with the south winds behind her. That is precisely the pattern followed by the ship in which the Ancient Mariner sailed. First,

> the STORM-BLAST came, and he
> Was tyrannous and strong:
> He struck with his o'ertaking wings,
> And chased us south along.

And then, after meeting the albatross,

> a good south wind sprung up behind;
> The Albatross did follow,
> And every day, for food or play,
> Came to the mariner's hollo!

Coleridge had never been to sea nor set eyes on an albatross and yet, miraculously, he understood that this pattern of coming and going, this ever-repeating gyre, living within the structure of the wind, is more intimately connected with the pattern of the bird's existence than any other element in its life. Albatrosses are the wind; their existence is inconceivable without it; and the weather systems are the metronomic masters of their lives. They have no landscape or seascape; they live in a windscape. Coleridge, intuitively extending that web of connections, understood that if the wind died, the bird would die. And the poem relies on the converse: if the bird dies, the wind will die:

> And I had done a hellish thing,
> And it would work 'em woe:
> For all averred, I had killed the bird
> That made the breeze to blow.
> Ah wretch! said they, the bird to slay,
> That made the breeze to blow!

Over the nearly three decades since his breakthrough, Henri Weimerskirch has been at the leading edge of seabird science in the world. He is now director of research at the French government Centre d'Etudes Biologiques at Chizé, 40 miles in from the Atlantic coast at La Rochelle, where he leads a team of six researchers and eighteen PhD and postdoc students. That team has proved to be one of the most powerful instruments ever assembled with which to understand these birds.

In 1993, they showed for the first time that albatrosses divide their fishing effort in the breeding season between short trips, to feed the chicks, and long trips, to feed themselves, revealing at the same time that females fish to the north of

Crozet, males to the south. A year later, by combining satellite transmitters, now down to little more than 2 ounces each, and special encased thermometers which could remain in the albatross's stomach and record changes in temperatures over a long fishing trip, they were able to give some shape to what was happening on those extraordinary foraging journeys. The albatrosses were eating an average of 4½ pounds of squid a day, or up to 2½ stone of squid in a single week-long fishing trip. Each time it ingested another lump of squid, its stomach temperature dropped by as much as 20 degrees. Sometimes the bird went fifteen hours between meals, at others chancing on a shoal and finding a mass of squid on which it repeatedly gorged. This pattern alone could explain the need for the enormous distances flown by the birds. Albatrosses cannot know where the squid is and so their life technique must be to quarter vast areas of sea for that patch in which they can eat their fill. This is at the heart of the bird's identity: vast ocean; enormous bird; 11 feet of wing; unthought-of journeys and distances and the speed needed to cover them. All are a reflection of the uncertainty of finding food.

Slowly more dimensions of the albatross's unseen life became apparent. In 1994, Weimerskirch attached a miniature water-temperature logger to the feet of six breeding wandering albatrosses, revealing the alternation of swimming or flying. If you can know whether the feet of the bird are wet or dry, you can know the pattern of its days. In 1999, he fitted three miniature instruments to his birds: one of those wet–dry activity-recorders, made in England; an American location-transmitter; and a miniature Finnish heart-monitor. For the heart-monitor to pick up the signal, Weimerskirch put two small electrodes of gold-plated safety pins under the albatross's skin. Together, this equipment weighed just over 2 ounces, 0.75 per cent of the weight of the bird.

The results of this experiment transformed the understanding of how seabirds live at sea. Previously, it had been thought that an albatross's flight across the ocean, recognized as far more energy efficient than any flapping bird, would nevertheless need the equivalent of three times the energy it would use when sitting at its nest. Not at all. The bird's heart rate was indeed low when at its nest, running at a steady 65 beats a minute. Easily the most stressful and energetic moments were when walking or taking off, either from sea or land, when its heartbeat rose to 230 beats a minute. Sometimes when sitting on the sea, the bird's heart would quite suddenly jump into action, starting to beat faster, as Weimerskirch guessed, when grabbing for a squid or fighting off a rival.

The most revelatory part of the results was what happened during flight. As soon as the albatrosses were airborne, their beating hearts steadied, the rate dropping quickly to about 80 beats a minute and within a couple of hours of continuous, gliding flight sinking still further so that as they cruised in the great gales that sweep around the world in these Roaring Forties, soaring over the peaks and into the giant troughs of the Southern Ocean swells, the albatross hearts were beating at the same rate they had maintained when back on the nests in the Crozet islands. The albatross was at home in the wind.

These giant birds hunt across 6 million square miles of ocean in their search for food, but at almost no cost to themselves. They almost never (less than 4 per cent of the time) encountered headwinds, no matter whether they set off north or south from their nest island. They always began by flying hard and fast in the westerlies, using them as side or tail winds, driving north-east or south-east to the hunting grounds. The faster they flew, up to 90 miles an hour, the calmer they became, making ever less effort the more the wind swung them around their ocean.

Albatrosses have developed an ingenious tendon lock in their shoulder, so that holding out their long and heavy wings involves no muscular effort. The wing simply hangs from their frame much as the whole bird hangs calmly in the wind. Distance means next to nothing to them, and for an albatross to travel a thousand miles in its chosen wind is no more work than the effort made by a man watching cricket in a deckchair for a summer afternoon.

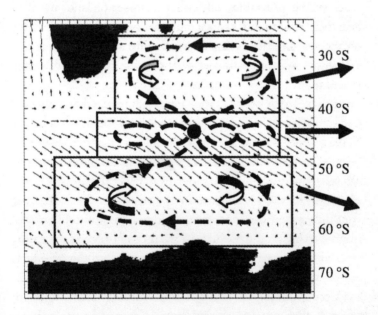

Constantly keeping the wind behind them, the Îles Crozet albatrosses swing on circular routes between Africa and the Antarctic

Nor is there anything random about their voyaging. Time after time, the Crozet albatrosses followed giant looping tracks, sliding off north-east or south-east downwind in the westerlies, or on a broad reach, with the wind coming over their shoulder, until they came to the latitudes either just north of Antarctica or just south of Madagascar in which the easterlies blow. There they turned west, turning as the wind turns, keeping it always behind them or to the side, pushing downwind, making up the longitude which would allow them finally, once they were again upwind of the Îles Crozet, to turn into the westerlies that would blow them home.

But that is the wrong word; these journeys are the albatrosses' home. The scale is astonishing: 7,000 or 8,000 miles in a few days, part of the normality of seabird life, living on a scale that hide-bound, terrestrial primates might find it difficult to imagine, but during which the birds' hearts stay calm, never more than when the ocean storms are roaring and crying around them.

In the first years of the twenty-first century, understanding of the albatrosses deepened with the developing technology. The Argos system had been able to deliver only ten or eleven locations a day. With birds travelling up to 90 miles an hour, that left enormous uncertainties over their routes and their behaviour. At the same time, early batteries could last little more than a month at best. That could tell the story of how the albatrosses behaved when incubating an egg or feeding a chick but could not hope to address their lives beyond the breeding season. With the miniaturization of GPS trackers and the development of longer-lasting batteries, more detail could be known for longer periods. The super-lightweight geolocator, of the kind which revealed the ocean journeys of the Eynhallow fulmars, was invented in the early 1990s. Its exceptionally low power demand meant that with it a bird's position could be recorded over years.

These technologies have reached the point where Weimerskirch can now claim that the wandering albatross is the best-known wild creature on earth. More than 600 albatrosses from Crozet and Kerguelen have been tracked. Every bird in Crozet has over generations been ringed when still a chick, all are known as individuals and the whole pattern of albatross life is starting to become clear.

When the chick, as big as its parents, leaves the nest it begins to circle around the globe. It has taken nearly nine months to reach this point since hatching and the parents will not try again for another fifteen months. They too will now have a 'sabbatical year' wandering around the Southern Ocean, sometimes circling Antarctica two or even three times, returning to food-rich places off the coast of Chile, New Zealand or New South Wales, foraging patches with which they have made themselves familiar over the decades.

Many of those young birds die. At least 25 per cent of them will be dead within two months of leaving the nest for the first time, and another 25 per cent before they are old enough to breed. One intriguing aspect of these figures, derived over many decades of albatross population studies, is that they are almost identical to the survival rates of human infants and children in pre-modern rural societies in Europe. About 25 per cent of all babies in seventeenth-century England died in the first month of life and another 20 per cent before they were five years old. The albatrosses, it turns out, are as good as people have usually been in looking after their young.

But survival is an education, and the young birds move up into relatively warm and deep oceanic waters, gradually extending their flight times and distances, so that by the end of the first year at sea they have flown on average more than 100,000 miles.

Once a bird is approaching its breeding age, it has a 95 per cent chance of surviving from one year to the next. It develops fixed habits, returning to its birthplace, always choosing the windiest, western end of islands and headlands to establish its nest, as take-off will always be easiest there. It finds a mate, performing elaborate high-wing-held dances to attract the other, sky-pointing and bill-clacking as it displays the great limbs on which their lives depend: 'Look, here, these are our future. Here are the wings on which we will live together for the rest of our lives.'

They stay with the same mate for decades. They learn to alternate with each other when looking after the chick, or out fishing for its food, so that in the middle years of their lives an albatross couple can expect to raise a chick nearly eight out of every ten times they try. But then age starts to tell. After they are about thirty, particularly for the males, their success at bringing up chicks starts to fall away. Even their habits diverge. When feeding their chicks, the females stay up in the warmer waters of the sub-tropics, but an old male will go ever further south on his foraging, deep into the Antarctic, his plumage whitening with every passing year until he becomes one of the great white purity birds that Melville saw on deck 'in waters hard upon the Antarctic seas'. By the time Melville saw its 'wondrous flutterings and throbbings', that bird had probably flown more than 5 million miles.

But this is not an unchanging world. The first impulse for the studies made by Henri Weimerskirch and others in the 1970s had been the fierce drop in albatross numbers they were witnessing; and, in a stream of scientific papers since then, response to change has been a constant theme. What happens to the albatross when the ocean warms around it?

With the great time-depth of their data, the French scientists on their supremely windy islands in the Southern Ocean have been able to track the birds' response to that changing world. Over the past fifty years, as the atmosphere and seas have warmed, and the temperature gradient between Equator and pole has steepened, the westerlies in the southern hemisphere have strengthened and shifted south towards the pole.

The albatrosses have responded to the fiercer winds by flying faster than before but not any further. As a result they have managed to complete the necessary tasks of feeding themselves and the chicks more quickly.

These changes have meant that the albatrosses have been able to spend more time at the nest. Most breeding failures tend to occur when the bird on the nest begins to starve after its partner has been out fishing for too long. As a result of those shorter-lasting fishing trips, breeding success has gone up. At the same time, in those fifty years, the albatrosses have put on weight. They are now, on average, 2 pounds heavier than they were in the 1980s, perhaps because they are better fed, perhaps because a heavier bird does better in high winds, able to use the energy of those winds more efficiently than a light bird that will be tossed to and fro. It may be that these stronger winds have selected heavier birds for survival.

One further and surprising benefit is that the female birds now fishing further south than before have removed themselves from the dangers of the longline Spanish and Japanese fishing boats going after tuna, which have been the cause of millions of albatross deaths over the last two decades. The use of these enormous longlines, each stretched 60 miles or more across the ocean, with tens of thousands of squid-baited hooks (the fishing fleet based on the French island of Réunion in the Indian Ocean sets 26 million hooks a year) had ballooned in the early 1990s after driftnets were discouraged and then finally banned in 1992 in order to protect dolphins and sea turtles. Longlines are a death trap for albatrosses, which try to steal the squid before the lines sink, get caught on the hooks and are carried down to their deaths.

Plans to set up exclusion zones in which longlining was banned were themselves scuppered by the satellite-tracking data: a map produced in 2003 showed albatrosses present and fishing across the whole of the southern hemisphere. The entire ocean could not be closed to the fishery. Conservationists developed so-called 'mitigation devices' – streamers that frightened the birds, lines to be set only at night when the birds are not flying or heavily weighted lines that sank too quickly for the birds to reach the bait – and reductions were made in the number of albatrosses that were being drowned each year.

Nevertheless, damage continues and longline fisheries are still now the greatest cause of albatross deaths. Because all legal boats now carry self-identifying transmitters, their location can be tracked by the scientists who are also tracking the albatrosses. Evidence from those combined signals shows that birds can detect and turn towards a boat from up to 15 miles away, stopping near it, following it or fishing in its area.

Before 1990, no albatrosses used to fish in the relatively shallow seas around Crozet. Now, at least a fifth of them do, only because they are attracted to the longline boats in the area. In some ways, the birds have become dependent on the boats that are killing them, particularly the pirate-boats which are unregistered and which have been decimating the birds. Dominique Filippi, a French engineer working for Sextant Technology in New Zealand, has come up with a new piece of equipment which is designed to turn this relationship on its head. The new tag, called an x-GPS, combines a GPS tracker with a radar detector. All fishing vessels, whether they are legally registered or not, keep their radar on to get advance warning of any fisheries protection vessels that, even in the wild conditions of the Southern Ocean, might be coming to look for them. The albatross wearing the x-GPS can transmit its own position to a satellite and at the same time, through the radar pulse, can detect and locate those illegal vessels, acting as monitor on the very fishing-boats which are likely to kill them. It is possible, in this way, that the albatrosses can become the source of their own salvation.

There is one final element in this extraordinary story: it may soon be possible, if one can express it in these terms, to understand what each albatross is like as a person. Samantha Patrick from Liverpool University and Henri Weimerskirch have tested the behaviour of a small sub-colony of black-browed albatrosses on Kerguelen. In one of the remotest and most beautifully named places on earth, the Cañon des Sourcils Noirs, the Canyon of the Black Eyebrows (which is the name in French of the black-browed albatrosses), a deep and narrow slot comes in from the sea on the south-eastern tip of the Presqu'île Jeanne d'Arc, with a small grey stony beach at its head. There are 200 nests of the black-browed albatrosses there. Samantha Patrick wanted to know which of the birds could be considered bold or shy, and whether the boldness or shyness they displayed on the nest had any relevance either to the way they behaved when out on their journeys across the ocean or to their success as breeding parents.

She tested the birds with what science likes to call a 'novel object'. If a bird is bold it will react fiercely towards it; if shy, it will shrink away. The novel object she chose was a large pink volleyball (2 feet in circumference), attached to the end of a 26-foot-long carbon-fibre fishing pole. Each bird had this appalling thing presented to them in turn while being filmed with a GoPro for a full minute before the ball was removed. Depending on whether they pecked the ball (bold), lunged at it (quite bold), said something vocal (quite shy) or looked away and 'snapped' – that is opening and closing its bill but not at the ball – (shy), Patrick scored them as shy ones or bold ones.

The population divided equally enough, and fifty-five birds were chosen as having strong identities, either clearly bold or clearly shy, and these birds were fitted with GPS loggers. When they went fishing, Patrick analysed the results. Would shy birds behave differently at sea from bold ones? Strikingly, they did. The bold albatrosses that had attacked the ball didn't bother to go far on their fishing trips but stayed just offshore. Shy birds, which had felt slightly overcome as the pink volleyball approached them, went fishing further away, in deeper waters, and in places where competition was not so fierce. These were the loners that liked a little isolation, away from the bold, aggressive and more competitive birds. Like many ornithologists, they wanted to find their own private corner and make the most of it.

The best places to go fishing, as it happens, are not immediately offshore at Kerguelen but up to 400 miles away, where the shallow shelf around the island drops off into deep ocean. If bold birds stay near and shy go far, which was Patrick's clear and repeated result, it may be, she thought, that the shy birds are also the careful birds, not simply mopping up the easy and possibly low-quality food near to hand, but taking the trouble to seek out the good spots and go fishing there. Bold in fact may be another name for clumsy, and shy only part of being careful.

From studies of shyness and boldness in bats and shrews, it also seems likely that shy is connected to clever. Shy birds probably rely more on memory than bold birds, holding in mind where the good spots are rather than relying on chance and sheer brute dominance to get the food they want. The psycho-structure of albatross life was starting to come clear: a world of bold, clumsy and relatively stupid bruisers interacting with shy, clever, careful and painstaking perfectionists, both of which had their place, each of which would respond differently to a changing world. It is probably true that the health of albatrossdom, even of seabirddom, relies on this social and psychological plasticity – resilience dependent on variability – if it is to face the future with hope.

This experiment was the first time anyone had looked at the shy–bold spectrum on land and combined it with fishing behaviour. But what effect did these qualities have on the albatross's fate? It seems as if there is a gender divide. The shyer male albatrosses always do better than bold males at bringing up chicks and bolder females do better than shyer females at least in those years when there are plenty of squid or fish in the sea. Why is that? It may be that bold male albatrosses, staying close to the colony in a crowded, macho environment, are always living in a state of high competitive tension which will exact its price, and maybe even age the birds faster and so reduce their effectiveness as parents. For the females, when there is plenty to eat, those which don't have to travel far will be better mothers. In years of dearth, the shy ones will do better, quietly sustaining themselves and their chicks out beyond the hubbub near the colony.

Samantha Patrick and Henri Weimerskirch also had a close look at the shyness and boldness of nearly 1,400 wandering albatrosses on Crozet. No pink volleyball this time, but watching how the birds responded to a person walking up to them. In the great white albatross, boldness was sex dependent: females were bolder than males and boldness in both sexes decreased with age. They found that nothing was more gentle than an aged wandering male, off most of the time on his giant forays down towards the ice of Antarctica, not bothering with much beyond his own adventuring. It was not clear to Patrick and Weimerskirch, though, whether the shyness of the old birds was because old birds had turned shy; or because bold birds live hard and fast, so that more of them had died young, leaving a population of careful, self-preservative albatrosses of the kind Melville had seen.

An Indian Yellow-nosed Albatross (*Thalassarche carteri*), east of New
South Wales

Birds that lived in big, dense colonies were shyer than those
that lived in little albatross villages and hamlets, separated
from each other by small valleys. Urban living, for albatrosses,
makes you modest. Only in villages does the full-on striding
dominant martinet appear. More impressively still, Patrick
and Weimerskirch showed that boldness was heritable. If you
had bold parents, you were likely to be bold yourself. If the
family was shy, you were also probably going to be shy. It was
the first time anyone in the world had shown that varieties of
seabird personality – whether taught at the nest or genetically
inherited remained unclear – may have had their origins in the
culture or the genes of the family transmitted from one generation to the next.

That is the conclusion to this long story of assiduous
research down in the windswept bleakness of the Terres
Australes Françaises: albatross life, for all its global heroics, is
like ours, a subtle interweave of dominance and dependence,

of needing others and excluding others, of reacting differently to different demands and situations, of qualities inherited, learned from the parents and understood over the course of a lifetime, oscillating between home and distance, partnership and singularity, inventiveness and long, long hours of dogged flight and searching over the emptiness of the stormiest ocean on earth. The songlines of the albatross are laced around the world; their *Umwelt* is one to which the only sane reaction is awe.

11

The Seabird's Cry

In the long dance of extinction and continuity, there are some birds that stand as symbols of persistence. The gyrfalcon is large, speckled and power-winged, the biggest falcon in the world. It is the size of a buzzard, living and breeding on the tundra shores of the Arctic Ocean, taking Arctic hares, little auks and ptarmigan as its prey in summer. In the winter it becomes a kind of seabird, out on the ice-bound wastes, like a figure from one of Coleridge's visions, the pale winter hawk on the frozen ocean preying by moonlight on razorbills and guillemots, alighting now and then on an iceberg, a secret seabird, each one quartering 25,000 square miles of blazing hostility for its kills.

The dry, cold conditions of the Arctic have allowed bird scientists from the Edward Grey Institute at Oxford to make an extraordinary discovery. The gyrfalcons remain loyal to the few sheltered nest sites that the cliffs of Greenland can provide and on those sites, usually on ledges under a rock overhang, the guano can build up to a depth of 5 feet or more. Over three summers, from 2002 to 2004, Kurt Burnham from Oxford took samples from the lowest level of the guano, where it touched the rock – carefully preserving the integrity of the nest – and had them carbon-dated. The results were revelatory: one gyrfalcon nest in Kangerlussuaq in western Greenland is between 2,360 and 2,740 years old. Others nearby were more than a thousand years old. The youngest they found had been occupied since 1350.

The gyrfalcons are not unique. The slightly larger northern guillemot called Brünnich's guillemot in Europe and the thick-billed murre in America has been found breeding on peat mosses where its ancestors were doing the same thing up to 3,800 years ago. The peat itself, 6 feet deep on the cliff edges of islands in Hudson's Bay and in Ungava Bay in northern Quebec, is the product of the birds having been there so long, fertilizing the sterile, frozen rock around them. In Antarctica, feathers and bones of Adélie penguins have been found on the edges of the Ross Sea that are at least 44,000 years old, although the ice ages that have come and gone between then and now mean those sites have not been permanently inhabited. The continuity record is held by the ethereally beautiful snow petrel, whose half-fossilized stomach oil has been found in the mountains of Queen Maud Land in Antarctica, the birds still nesting where they were nesting 34,000 years ago.

In the warmth and damp of the temperate and tropical worlds, where rot is universal, these materials would not have survived. It may well be that the birds we know and love have been here as long as their cousins to north and south, but we can find no evidence. Those polar stories are the lessons we need to hear. The oldest surviving physical text in the Western canon – an *Iliad* from a grave in Egypt – is younger than the gyrfalcon's nest in Kangerlussuaq. The great cave paintings of the palaeolithic are not as old as that oil from the snow petrel's stomach. The penguins were doing what they do now well before humankind was in Europe or the Americas. It is a time-difference that allows us to see ourselves as we are: vagrants in the world, Calibans, gazing at the creatures afloat above us.

I know from experience how blind one can be to deep and threatening change. Our natural assumption is that we live in constancy. The kittiwakes on the Shiants are now 80 per cent down in number from the birds I used to see in the 1960s. But if I didn't know that, I am not sure I would be aware of it. The colonies that are now gone were on some sea stacks on the far side of a mile or so of difficult, tide-churned water from the main island. In a small boat, it was always a challenge to cross, and those kittiwakes lived in their own little Shiant version of remoteness. I hardly ever went out there to see them.

The colony I know best, a core site, noisy and mesmerizing on the columns of a small dark inlet on one of the bigger of the Shiants islands, is only half an hour's walk from the bothy and audible for hundreds of yards before you reach it. Things have nearly always seemed to be all right there. Photographs I have of that cliff in 1970 almost exactly match its summer appearance now, the same perched nests, the same stained curtains of guano beneath them. There have been years in which the kittiwakes have decided to skip breeding – a wide-spread seabird habit – but apart from those blips, I have assumed that the world was unchanging and, somehow, with-out science, or precise record-taking, I failed to detect the decline. An 80 per cent drop in fifty years would need on average no more than a 3 per cent reduction each year, unno-ticeable to the casual eye. My baseline of normal has followed behind me like a dog, wagging its tail, assuring me everything is all right.

Compare like with like, though, and it is clear that the loss is real. Short pieces of film survive of me and my father taken in 1971, him fifty-two, me twelve, walking together along the clifftops of one of the Shiants, talking and then sitting side by side, him holding me back from the cliff edge, to watch the fulmars swing and wheel below us. The air is full of them in the film, never a moment of inaction, fifty, sixty birds in their stiff-winged perfection circling in the updraughts, playing with the wind, filling their life-space not like children in a playground, but graceful, a ballroom of birds 300 feet deep and a few hundred yards long, beautiful for the completeness of it, those mottled grey wings sweeping and re-sweeping across the air below us.

It is not like that now. The fulmars have largely gone and if I sit there today, on the same knoll, looking down at the suck and surge of the Minch on the rocks below me, only four or five fulmars are there, each of them as entrancing as their grandmothers or great-grandmothers had been, but more lonely than that, with too much air between them, as touching and vulnerable as the last of the travellers in a station at midnight.

*   *   *

Over the last sixty years, the world population of seabirds has dropped by over two-thirds. One-third of all seabird species is now threatened with extinction. Half of them are known or thought to be in decline. Some petrels, terns and cormorants have been reduced to less than 5 per cent of the numbers that were alive in 1950. Albatrosses and shearwaters, frigatebirds, pelicans and penguins have all suffered deep body-blows. Some bird families – the gannets and boobies, some gulls and storm petrels – have managed to keep their numbers up or even to increase them slightly, but overall the picture is a decline in seabird numbers of about 70 per cent in six decades. A 2012 study by the great British Antarctic bird scientist John Croxall and others found that seabirds were in more danger than any other class of vertebrate. Those seabirds whose numbers are even roughly known have dropped from about 300 million in

1950 to about 100 million in 2010. Individual species such as the Guanay cormorant on the Pacific coast of Peru have crashed from more than 20 million to 2 million in the same time. Scale up the decline to the global seabird population as a whole and the figure you get is one billion fewer seabirds now than in 1950. The graph trends to zero by about 2060.

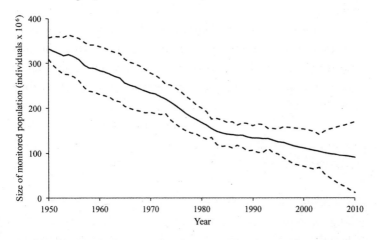

The long-term trend in those seabirds that have been repeatedly counted over the last sixty years is inexorably downwards

That won't happen; the seabird phenomenon is too complex, with too many unknowns and too many local variations for any single figure to embrace its future, but there can be no doubt that we have brought this disaster on ourselves: through overfishing; by the massive accidental catching of birds in fishing gear; by their deliberate destruction; by introducing rats, cats, dogs, pigs, goats, rabbits and cattle to the breeding places of birds which were defenceless against them because they had not evolved with them (many Pacific birds literally cannot smell a rat); through pollution by oil, metals, plastics and other toxins; by the destruction of nesting sites by human development; and through the multiple effects of climate change and the acidification of the sea.

Of the ten birds that appear in this book, seven are in decline, at least in part of their range. But complexity and variation is inseparable from their future. In the Pacific, fulmars are actually increasing and, with 20 million individuals in the northern hemisphere, plus another 4 million of their southern cousins in the oceans around Antarctica, they will be here for a while yet. But there are worrying signs: they are being caught by Spanish longliners; they are drowning in gill-nets, which hang in the sea and are designed to trap fish by the gills but are just as effective at catching birds; and they are being killed under the heavy cables behind trawlers where they gather to feed on offal. In European seas, fulmar numbers have declined by 40 per cent over the last thirty years. They are due to crash by the same again in the next fifty, perhaps a return to the numbers in the Atlantic before the great boom in industrial fishing, perhaps a more disturbing descent into rarity.

It is the same story for the Atlantic puffins, now with about 11 million birds alive but under frightening pressures from climate change and the disappearance of their prey. They too will be 80 per cent down by the second half of this century and will be a rare sight anywhere south of the northernmost islands in Scotland. The tufted and horned puffins, of which there are about 5 million in the Pacific, are also declining. Their cousins, the guillemots and razorbills, are suffering from the oceano-graphic changes that are affecting the puffins, and although there are still enormous numbers (about a million razorbills, 18 million guillemots and 22 million Brünnich's guillemots) there have been sharp losses in Iceland, one of the headquar-ters of auk life, with big colonies losing almost half their birds in the three or four years after 2005. There are now some 20 million kittiwakes strung around the northern hemisphere, but in places the population has halved in number since 1983, further victims of climate change, windfarms, oil spills and industrialized fishing taking too great a portion of their prey. Research has shown that seabirds do not suffer if fishing fleets remove two-thirds of the fish that are usually available. But take any more than that, as fishermen have often done in the last 100 years, and birds begin to die.

The great black-backed gulls are holding their own, the 2 million lesser black-backed gulls are actually on the increase, but the herring gull is going, dropping 30 per cent in the last forty years, many of them getting caught in fishing gear. Cormorants are doing well, often by colonizing inland waters, but most shags in the world are on the slide. No one knows what the trends in Europe are for the 1 million Manx shear-waters. They are suffering in North America and other shearwaters around the world are nearly all decreasing. The gannet is the great North Atlantic exception, a powerful, wide-ranging generalist, not stuck in any kind of ecological cul-de-sac but able to fly great distances and catch almost anything when it gets there. Gannets are booming, with at least 1.8 million birds presiding over what the Anglo-Saxons liked to call the 'gannet bath' of their cold northern ocean.

The anxieties are running highest with the albatrosses and petrels of the Southern Ocean. The rarest of all, the Amsterdam albatross, a beautiful dark-backed creature, a close relative of the wandering albatross but breeding only on a single marsh on the tiny Amsterdam Island in the southern Indian Ocean, is down to twenty-six breeding pairs. This is better than its worst moment in the 1980s, when five breeding pairs were in exist-ence, but the species is still under threat of extinction. The death of six adults per year would bring it to an end. There are 20,000 wandering albatrosses, but their breeding success is no longer reliable and overall numbers are still dropping, despite some highly effective mitigation measures on longliners around South Georgia, where the albatross deaths have decreased by 99 per cent in ten years. Other albatross chicks on Gough Island in the South Atlantic are being eaten by giant mice, which are threatening them with extinction. Overall, between 160,000 and 320,000 seabirds are being killed each year by longlining vessels, mostly from Japan and Taiwan. Streamers on the line and an insistence on setting them at night avoids much of the mortality, but on moonlight nights the albatrosses are still dying and, anyway, illegal and unregulated fishing is rife.

Efforts are being made all over the world to counteract these agents of the apocalypse, to reduce the accidental by-catch of birds, to eradicate the predators (more than 300 islands have now been cleared, with a 90 per cent success rate) and to limit or even ban the shooting of seabirds for food or fun (even if Newfoundlanders and Greenlanders still kill up to 400,000 guillemots each year), but those measures have not yet halted the overall decline. The threat posed by gillnets, the loss of breeding habitat, the ever-growing human presence in places which were previously hard to reach and the rippling, multi-level interactions of a warming sea and the layers of life within it are all still bowling the seabirds towards calamity.

In the summer of 2015, I went to a colony in the North Atlantic where a version of the end has already come. Heimaey is the largest of the Westman Islands off the south coast of Iceland, its streets now rich with cod money, but, underlying its neat houses and surrounding its careful harbour, Heimaey is still a sharp and jagged lump of volcanic rock, recently emerged from the Mid-Atlantic Ridge and ready to erupt some more. Erpur Hansen, a seabird scientist based in the Westmans, took me one evening to a puffin colony he had long studied, on the peninsula called Stórhöfði, almost on the

southern tip of Heimaey. The long-fretted skyline of the mainland of Iceland lay to the north. To the south of us, the other
Westman islands were scattered across the evening ocean like
objects set out on a library table, neat and decorative, each of
them bright lush green with their skin of grass, and each with
a tiny white house perched like a kittiwake's nest above the
cliffs, shelters for members of the puffin-hunting clubs that
like to claim each diminutive island as their own.

Diddi Sigursteinn wishing good luck to one of the few remaining
puffins on Alsey, Westman Islands, Iceland

Erpur, with a deep and saturnine air about him, dressed in
black from head to foot, unshaven, serious, weary, stopped the
car and we walked a few hundred yards to a shallow slope high
above the sea, dished into the hillside like a Greek theatre.
'Here we are,' he said. I saw nothing at first beyond a wooden
birders' hide below us. Then, more closely, thousands and
thousands of burrow entrances stippled into the hillside like
the flecks in tweed. This was a puffin colony with maybe 2,000

burrows in it. You could have expected five or six thousand birds to be there. But this evening, there was none. I looked hard and then out to sea, and found with my binoculars about twenty puffins, circling forlornly in the ocean of air.

On any given day, a puffin colony will either be full of life or empty, crowded with birds or every one out fishing. They come and go like that. Had Erpur and I simply arrived at a moment when all the birds were out to sea? No, he said. There's a problem with food. In the 1990s this would have been packed with birds. But that's now history. This is the thirteenth year in a row we have seen a breeding failure of lesser or greater extent. The birds that should be here now were never born. Now we are seeing only the remains of the old adult population, of the adults that were here before the breeding failure began. Their numbers are going down slowly, year by year. And they are not being replaced. In his mist nets, set up across the colony, Hansen is lucky if he catches five or ten birds in two hours. On the Shiants, that figure might usually be 200 or more.

Visiting Iceland's puffin islands in the early days of the failure, Erpur would come to a colony in which all the chicks were dead, most of them about the same age and all dead within a few days. On one island, Papey, there were 130,000 dead chicks. 'There was a sweet smell in the air, a death smell. And that's not easy.'

Sometimes, the adults abandoned the eggs before they hatched. At others, the chicks were abandoned unfed. 'It is part of the calculation of seabird life: if the adults are not feeding themselves, they would rather give up the breeding attempt and return another year. They're not sacrificing their survival for one attempt, which is going to fail, because if they are hungry the chick will probably not survive anyway.'

As we spoke, another car pulled up and a family of two

young teenage girls and their parents walked along the path towards the birders' hide. They had come to see the puffins. Yes, Erpur said, I guess all the guidebooks are still saying this is a place to see puffins. The Westmans were once the largest puffin colony in the world.

Vast and equivalent failures to feed puffin chicks have been logged in the Norwegian Sea. Mass death occurred there year after year from the 1970s onwards and the population dropped from almost 5 million puffins to about 1.5 million in forty years. 'If you want melancholy,' Erpur said, 'go and talk to the Norwegians.' It is a pattern that is appearing all over the North Atlantic. Some populations have been holding up, in Britain and Canada, but projections for global puffin populations now envisage a 50–80 per cent decline over the next sixty-five years.

Although many uncertainties lie ahead, what is happening is not complicated. Puffins are expensive to maintain and will die if they do not eat in three days. If fish are not near to hand, they can burn half their body-weight in energy every day going in search of it. Emily Scragg's GPS trackers on Shiant birds had shown fishing trips averaging about 12 miles, out into the sea lochs of Lewis and Harris and the shallow seas of the north Minch. The desperate birds on Stórhöfði were travelling at least 40 miles and occasionally as much as 80 each way in search of fish. The result was that those adults were spending nearly all the time fishing to pay for their own journeys and so the food supply for the chicks was radically reduced. Only a tiny proportion of all the fish caught was coming back to the chicks. And the further away the parents had to fly, the more likely it was that the chick would starve. Chick survival depends on local fish.

Why were the fish not available? In Norway, where the puffins were feeding on young herring that drifted up through

the islands off the Norwegian coast, the problem had begun
with overfishing but had nevertheless persisted even when the
fishing effort had been reduced. Overfishing for sandeels was
certainly implicated in kittiwake declines in the North Sea
during the 1990s, but not here with the puffins. In Iceland, no
one has ever fished for the sandeels, but they had disappeared
all the same. Something else was changing.

The steps in the argument, assembled by a host of sea scientists from all over Europe across many years, go as follows. The
sandeels have relied as their main food source on a small translucent planktonic shrimp called the *Calanus finmarchicus*. This
little shrimp, about one-eighth or one-sixteenth of an inch
long, exists in the cold North Atlantic in such vast numbers
that the ocean could be thought of as an enormous bowl of
Calanus broth. Fish, bigger shrimps and whales all eat them
because, grazing on still smaller animals and floating phytoplankton, *Calanus finmarchicus* concentrates large amounts of
protein and omega-3 fatty acids, principally so that it can last
out the long, cold and almost foodless northern winters,
during which it hibernates, living on its own fat reserves. Only
when the spring returns and the phytoplankton blooms in the
warming, well-lit summer ocean does *Calanus finmarchicus* feed
again.

That hibernating phase, lying at the bottom of the ocean and
requiring large fat reserves, is the critical link in the well-being
of the ocean food web on which these birds rely. *Calanus* fats
feed sandeels which feed puffins. But *Calanus finmarchicus* has
been disappearing from the North Atlantic over the last twenty
years, to be replaced by its southern cousin, *Calanus helgolandicus*, which habitually lives in warmer waters of the continental
shelf and does not need to hibernate. Instead, it grazes all year.
The number of *Calanus finmarchicus* has dropped by 70 per cent
since the 1960s. But the implications for the food web are

radical. *Calanus helgolandicus* does not require such large fat reserves. Because it goes on eating, it can stay slim all year. This is a disaster for the fish that until now have relied on *Calanus finmarchicus* for their food supply, simply because for every shrimp they now eat, they get far less fat than they would if grazing on *Calanus finmarchicus*.

But why has *finmarchicus* been replaced by *helgolandicus*? For one simple reason: the sea has warmed up south of Iceland. *Finmarchicus*'s habit of hibernating has become an irrelevance in a warmer ocean; much more suitable to be a steady, all-year-round grazer. The result is a decline in the production of sustaining fats by the plankton at the base of the food pyramid. The sandeels that rely on those fat plankton for their own fatness are now either thin or find it difficult to survive the winter themselves. If the winter seas are warmer than before, their own metabolism runs faster, they get through their stored-up fat supplies more quickly and they die before the fat can be replenished in the spring. At the same time, bigger sandeel-eating fish, the mackerel and blue whiting, are also thriving in the warmer seas and so the sandeels are caught from both ends: not enough food for them to eat and too many predators eating them. The impact of the new mackerel shoals also represents major competition for the seabirds. It is a straightforward cascade: warm seas, thin new species of plankton, thin or absent sandeels, voracious mackerel, hungry seabirds, dead chicks.

The change is dependent on the circulation of waters in the North Atlantic. Southern Iceland is on the boundary between the North Atlantic Current – a stream of warm sub-tropical water, flowing up the Eastern Atlantic past the British Isles from the south – and a circulating pool of cold oceanic water in the centre of the ocean, the sub-polar gyre. In a cool world, the gyre is large and energetic, keeping the warm waters well to the east of Iceland. When that gyre loses energy, in a generally warmer world, the boundary between the two moves west. Even a mild fluctuation in the regime will bring sudden effects. In the mid-1990s, that boundary did move west, perhaps stimulated by overall global warming, and flooded the deep ocean south of Iceland with a river of much warmer water. A wobble in the great circular gyre of the North Atlantic banished the *Calanus finmarchicus*, got rid of the sandeels, drove the adult puffins to enormous, self-destructive efforts of parenthood and starved the chicks. A warmer ocean had killed the puffins.

The puffin-hunters of the Westman Islands still go each year to their small huts on the outlying islands, for dinner and a glass of wine or two, but there are hardly any birds to kill. Instead, the photo-albums are brought out, of life as it used to be, with skies as full of puffins as a Scottish evening full of midges, with vast piles of dead birds on the island hillsides, more than 2,000 sometimes killed by the party in one day, picked out of the sky with the fleyg net, the neck twisted, the body set aside, the orange feet and hooded eyes flickering as the puffin expires. But the hunters do not give up their summer visits to these lonely outposts, largely because they take their young sons with them and want to keep the tradition alive, not because it is viable now – they recognize it is not, never killing those few adults still bringing sandeels into the burrows – but because they have a belief that this terrible disappearance is no more than the latest turn in the cycle of nature.

I spent the night with the hunters on the little Westman off-island of Alsey, drinking and laughing with them in their hut, looking out from its windows at the rest of the archipelago in the long, calm, pearly summer evening, with the storm petrels hidden in the burrows churring and burbling around us, as the surf pulled and clawed at the rocks far below. It was beautiful, full of bonhomie and charming men, fishermen, lawyers, bank managers and shopkeepers from Heimaey. There was no hint of final catastrophe. The Alsey men were sure the puffins would return. Fifteen years ago, as one of the hunters, Diddi Sigursteinn, told me, it had been 'completely white here when you arrived. Of course, there was green grass but it looked more white than green, white with their white chests.' There were probably 700,000 puffins on Alsey until about 2004. Now the Westman local government has restricted puffin-catching to three days each summer, but even that is too much. Years go by in which the Alsey men do not catch a single bird.

Erpur Hansen has brought together the records of the Westman puffin-hunters with sea-surface temperatures over the last 100 years and has found a striking correlation. Numbers of puffins caught and killed have dropped away three times in the century: once in the 1880s and 1890s, once in the 1930s and 1940s and now. Each time, on something like a sixty- or seventy-year cycle, the sea had been warmer for a while. It is a multi-decadal oscillating cycle of which the puffins are a clear barometer. Hansen reckons that by about 2020, the cycle should reverse again into a cold phase. If he is right, the plankton regime will shift back towards the more fatty *finmarchicus*, the mackerel will again slide south and the puffins will return to southern Iceland.

Neither he nor other scientists are as sure as the hunters that this phenomenon is merely the turnings of a natural cycle. There is too much evidence of unprecedented warming in the world ocean for the current seabird crash to be blamed entirely on normal oscillations in the sea. Those oscillations may be occurring, but against the background of a far larger change. Long records of sea temperature around South Georgia in the South Atlantic now show that the ocean there is 2.3 degrees warmer than it was eighty years ago. The westerlies in the southern hemisphere are reliably 15–20 per cent stronger than they were forty years ago. The production of krill, the tiny *Calanus*-equivalent crustaceans on which the whole southern ocean ecosystem relies and which thrive in the cold, is now between 38 and 81 per cent lower than it was in the mid-1970s. Marine organisms are, on average, moving polewards by about 40 miles a decade. At the same time, for each decade, all the signs of spring in the oceans are appearing about four days earlier.

Deep perturbations to the system are evident wherever you look, but nothing is simple and everywhere counter-intuitive effects make themselves apparent. As the auks were suffering from the lack of sandeels in the warmer seas, Icelandic eider ducks were doing better with the slightly warmer summers. In the West Indies, global warming has meant that Caribbean shearwaters are raising more chicks than before, as their prey fish are growing fatter on the increased growth of phytoplankton. The blindingly rare Amsterdam albatross seems to thrive on the stronger winds of a warmer world, even though avian cholera is now also more prevalent in the warmer air and is killing its chicks. Melting glaciers in the high Arctic now create richer feeding grounds for little auks where the cold meltwater reaches the sea. Global warming has, paradoxically, created more sea ice in parts of Antarctica, miserable for certain kinds of penguin whose breeding colonies are now too far from open water. And in the Arctic, seabirds have now been identified as a source of global warming themselves, pumping methane and nitrous oxide into the atmosphere above their breeding colonies.

Global warming is troubling the ocean system as if it were an infection of its inner workings. Its deep changes to ocean currents, altering the kinds of fish in the sea and their distance from breeding colonies, when combined with unsustainable fishing methods and pollution of many kinds, will continue to do damage to the birds. With the puffins of Norway and the Westmans, resilience has been stretched beyond breaking point. The decline of guillemots in the Baltic is a measure of the drop in the number of sprats in that sea, probably damaged by fishing and by climate change. In what should be the cold, rich waters of the Benguela Current off South-west Africa, the fate of the birds has become the measure of a general crisis. Penguins, Cape cormorants and gannets have all virtually disappeared, the result of warming waters and the utter destruction by fishermen of the sardine and anchovy stocks which had previously thrived in those seas.

This is a resetting of life. The fossils will be different in the future. And in this enormous process, which we are now in the midst of, there will be winners and losers. Many bird lineages – early gulls, albatrosses, shearwaters, cormorants and penguins – survived the mass extinction 66 million years ago, when three-quarters of all other life disappeared, including most of the plankton. No doubt, in our present catastrophe, that will happen again. The great generalists and the great travellers, as long as we have not destroyed them first, may have the better chance to outlast the mass extinction. Those that are caught within narrow niches, either in their life-habits or in their body-form, will be more vulnerable. But there is no doubt: this is one of the ages of loss.

There may be some hope in the resilience of seabirds and in what biologists call 'plasticity' in their behaviour. A population of birds has within it enough variation and individuality for them to respond to change. Everything in this book that has described the intelligence and idiosyncrasies of different birds must suggest that their lives go beyond the purely mechanistic. Seabirds are often described as the 'barometers' of the ocean, the visible measures of hidden processes far beneath, but that is scarcely an adequate term. Instead, as Christophe Barbraud, a colleague of Henri Weimerskirch at Chizé, has written, they are more 'sentinels' than barometers, not merely the recorders of disaster but the animals keeping watch at the gates of extinction.

In the 1970s, the longline tuna fisheries of the Southern Ocean had been killing many of the wandering albatrosses out fishing from the Îles Crozet. The population had sunk from 500 pairs in the 1960s to 200 pairs in the 1980s. 'But', as Christophe Barbraud has said, 'what is surprising is the increase in the size of the population that followed, even while the fishery went on as before. By 2003, there were 400 pairs.' The explanation is partly Darwinian and partly based on the ability of the albatrosses to learn. 'It is the law of selection,' Barbraud has written. When animals are harvested in the wild, either deliberately or accidentally, the very process of harvesting will lead to the extinction of those traits which the harvest selects. If hunters kill big-horned sheep, small-horned sheep will survive. If fishermen catch only big fish, small fish will survive.

With the Crozet albatrosses, the tuna fishermen were, inadvertently, killing those birds which liked to pick squid from their hooks. The by-catch was removing the bold and daring birds from the population. The result was that the remaining albatrosses were the meeker ones, an inheritable personality trait, so that by 2000 the Crozet population had been stripped of its foolhardy genes. Those that were left avoided the long-lining boats and the numbers started to rise. But was this change merely Darwinian selection? Or were the albatrosses learning not to die at the tuna boats? Barbraud does not commit. 'Both are possible,' he says.

Here, maybe, is some cause for hope. Seabirds are not necessarily passive victims of the giant changes occurring to their world. They respond, they fight, they learn, and they have within and between them an adaptability and a resourcefulness which the very processes of destruction can bring to the fore. The expanding gannet populations of the North-east Atlantic or the cormorants flooding into freshwater lakes and rivers, surviving on their adaptability, are signals of future well-being.

There are no grounds for complacency. The great extinction is going on every day and the rate of change in the nature of the oceans is almost certainly too rapid for many of the inbuilt resilience mechanisms to cope. Most people who understand the question, and who have devoted decades of their lives to these animals, think that most of the birds will suffer. Their ranges will shrink. There will be many local losses and several global extinctions. Birds that are now common will become rare. Populations will shrivel. The grand cry of a seabird colony, rolling in its clamour around the bays and headlands of high latitudes, will become a memory, its absence unnoticed because people will not miss what is not there.

What is to be done? Only all that can be done. We must not preside over the ending of these marvels, or have their absence as our memorial. Most of the practicalities are obvious: there needs to be a much wider understanding of the state of the world's seabirds. Their breeding places need to be protected from people and their predatory animals. Much tighter controls must be imposed on all kinds of fishing vessels so that fewer birds are killed at sea. Because the death-rate of adult birds in the winter has more influence on seabird populations than any other single factor, their winter feeding grounds need to be understood, mapped and protected, particularly the great oceanic upwellings off the continents. Finally, and most intractably, the rate at which we are changing the atmosphere and the ocean, both its temperature and its acidity, needs to be brought under control.

A few years ago, I went for a walk at Sissinghurst in Kent, a place I have loved all my life, and took with me a friend, Claire Spottiswoode, an ornithologist now working at the universities in Cape Town and Cambridge, but who was brought up in Africa and has spent her life documenting the birds of that still-enriched continent. After a while in the heavy summer woods and fields, I turned to her and asked what she thought of this place, expecting her to love its deep and rooted beauty, the shadows of its oakwoods, the glow and burnish of its summer meadows. 'There is nothing here,' she said. 'Where is everything? Where are the animals? Where are the birds? It is empty. Everything has gone.'

I had never seen that, never felt the absence, nor understood the degree to which we had removed the life from the country in which I had lived for so long. Where were the birds? Long gone, destroyed by chemical farming. And the animals, most of them exterminated thousands of years before that. Nor had I seen the connection between that absence and everything that had first drawn me to the Shiants and its birds. The Shiants were a fragment of the undiminished world. That is where I had been given a glimpse of precisely what was absent from every other place I had ever known. The seabirds in their colonies were and are a last bastion of wholeness, insulated from our destruction by the enveloping protection of an ocean on which they were at home and where, thankfully, we never can be.

In one minuscule way I have been involved in doing something useful. Ever since I have known the Shiants, which my father gave to me in 1978 when I was twenty-one, the islands have been overrun by the ship rats. They have been there since they came off ancient wrecks, probably in the eighteenth century. In 2009, a study by government nature scientists and the Royal Society for the Protection of Birds (RSPB) concluded that, of all the bird islands in Britain, the Shiants would benefit most from having the rats removed. It was something I had dreamed of all my life. Money was raised, talks were given and dinners held, so that finally, over the winter of 2015–16, a team of specialists, largely paid for by the European Union and Scottish Natural Heritage, arrived on the islands to do the work.

For year after year, the rats had eaten the eggs and chicks of the more vulnerable birds. Now, for five brutal months, dark for eighteen hours a day in the pit of winter, the members of the rat team were going to remove that burden. All winter, they slogged across hillsides and dangled off ropes set down the cliffs. One night, they felt their Portakabin, anchored to the ground on top of one of the islands with giant hawsers attached to concrete blocks, sliding cliffwards under the hurricane-force winds. It had moved about 8 feet before they managed to leap out into the storm where they cowered all night before rescue came.

Bait was repeatedly laid for the 3,000-odd rats. Every corner of all the islands had to be covered and every last rat killed. If one pregnant female was missed, the whole exercise would have failed. Most of the rats died within a few weeks. One or two hung on, nibbling the poisoned chocolate and peanut butter with which the team was tempting them. By the time the spring came, no bait in any of the 1,800 stations set out across the islands was showing any signs of being taken. The rats had gone and the ground-nesting birds, above all the vulnerable storm petrels and the precious Manx shearwaters, could return to the islands on which they had been unable to breed for about 300 years.

It is, on an ocean scale, the slightest of gestures, an attempt to ensure in one small corner of the Atlantic that the birds will have one less hurdle to jump, one less factor driving them to extinction and one extra fragile defence against the forces of destruction. The hope is that the Shiants might become a refuge and a gene-source for other less protected places, so that from here the seabirds will proliferate across wide swathes of the North-east Atlantic.

A young Hebridean wren on the Shiants

I hadn't predicted what would happen that spring. Everybody said it would take years for any change to become apparent. The life-structure of seabirds means nothing changes quickly. All resilience is founded on slow and stable responses. But no one had considered the landbirds. I was there with my family that summer. Every day, we woke and looked out through the door of the bothy, and there all around us, for the first time, standing on the fence-posts and the grey walls of the old buildings, were families of wrens and wheatears. They were nesting in the cracks and crevices of the old houses, able to raise their multiple young because for the first time the rats were not destroying them. A light, buoyant bubble of bird life filled the air around the house. They were here unsummoned and of their own accord. They were not seabirds and what we were listening to was scarcely a cry, more a miniature rivulet of song, trickling out into the air above the surf, but they were something new, or at least something restored: little dancing apostles of the future, springing from one wire fence and one lichened stone to the next. Obedience to gravity? Not here, not for the moment.

Notes

Introduction

2 The greatest sin: Simone Weil, *La Pesanteur et la grâce*, Paris: Plon, 1947
(1988), p. 11. *Pesanteur* could also be translated as heaviness, viscosity or
stuffiness. Heaney quoted it in the first of his 1989 Oxford lectures, printed
in *The Redress of Poetry*, Faber & Faber, 1995, p. 3

2 'wi' its qualms': in Kevin MacNeil (ed.), *These Islands, We Sing: An Anthology of
Scottish Islands Poetry*, Edinburgh: Polygon, 2011, p. 9

3 'the earth earthy': from 1 Corinthians 15.47: 'the first man is of the earth,
earthy'.

3 'the windy light?': in Seamus Heaney, *Seeing Things*, Faber & Faber, 1991,
p. 78; Helen Vendler, 'Seamus Heaney's Invisibles', *Harvard Review* 10
(Spring 1996): 37–47

3–4 'songs you love': adapted from the translation in T. P. Dunning and
A. J. Bliss, *The Wanderer*, New York: Appleton-Century-Crofts, 1969

4 'no mead-drink': adapted from the translation in Charles Carpenter Fries,
The Seafarer: One of the Oldest English Poems, G. Wahr, 1925

5 'then they close': William Wordsworth, *Prelude* XI. 336–7

5 *inscendence*: quoted in Andrew C. Revkin, 'Thomas Berry, Writer and
Lecturer with a Mission for Mankind', *New York Times*, 3 June 2009

9 to catch them: Anthony Gaston, *Seabirds: A Natural History*, A & C Black,
2004, p. 177

11 'with one's skin': Arthur Koestler, *The Invisible Writing*, Random House,
2005, p. 432; for the origins of the 'oceanic feeling' embraced by Koestler
see William B. Parsons, *The Enigma of the Oceanic Feeling: Revisioning the
Psychoanalytic Theory of Mysticism*, New York: Oxford University Press, 1999,
p. 19

11 'pre-eminent, alone': Robert Browning, *Fifine at the Fair*, 1872, XXIX

12 '400 harp seals': Todd J. Kristensen, 'Seasonal Bird Exploitation by Recent
Indian and Beothuk Hunter-Gatherers of Newfoundland', *Canadian Journal
of Archaeology / Journal Canadien d'Archéologie* 35.2 (2011): 296

12 on the wing: Marine Grandgeorge et al., 'Resilience of the British and Irish
Seabird Community in the Twentieth Century', *Aquatic Biology* 4 (Dec.
2008): 187–99

12 in Nova Scotia: Brett C. Millier, *Elizabeth Bishop: Life and the Memory of It*,
Berkeley: University of California Press, 1993, pp. 181–2

12 'flowing, and flown': 'At the Fishhouses', from E. Bishop, *The Complete Poems
1927–1979*, New York: Farrar Straus & Giroux, 1979, online at
https://www.poets.org/poetsorg/poem/fishhouses

13 'depend on it?': Elizabeth Bishop and Robert Lowell, *Words in Air: The
Complete Correspondence between Elizabeth Bishop and Robert Lowell*, ed. Thomas
Travisano and Saskia Hamilton, New York: Farrar, Straus & Giroux, 2010,
p. 553

13 was its bowstring: Carl Safina, 'Wings of the Albatross', *National Geographic*,
December 2007, online at http://ngm.nationalgeographic.com/print/
2007/12/albatross/safina-text

15 about two-thirds: Michelle Paleczny et al., 'Population Trend of the World's
Monitored Seabirds, 1950–2010', *PLoS ONE* 10.6 (2015): e0129342

15 a brain tumour: Tom Lubbock, *Until Further Notice, I Am Alive*, Granta Books,
2012, entry for 25 October 2008

16 'for much conversation': Nicolas Fontaine, *Mémoires pour servir à l'histoire de
Port-Royal*, Utrecht, 1736, vol. 2, pp. 52–3

17 modern seabird studies: see Carlo Brentari, *Jakob von Uexküll*, Biosemiotics
9, Dordrecht: Springer, 2015; T. Rüting, 'History and Significance of Jakob
von Uexküll and of his Institute in Hamburg', *Sign Systems Studies* 32.1–2
(2004): 35–72

19 'all available worlds': Frans de Waal, *Are We Smart Enough to Know How Smart
Animals Are?*, London and New York: W.W. Norton, 2016, p. 8

20 'peaks of specialization': ibid., p. 12

22 in their stomachs: Chris Wilcox, Erik van Sebille and Britta Denise
Hardesty, 'Threat of Plastic Pollution to Seabirds is Global, Pervasive, and
Increasing', *PNAS* 112.38 (2015): 11899–904; http://www.sciencemag.
org/news/2015/08/nearly-every-seabird-may-be-eating-plastic-2050

1. Fulmar

25 Do fulmars play?: N. J. Emery and N. S. Clayton, 'Do Birds Have the
Capacity for Fun?', *Current Biology* 25.1 (Jan. 2015): R16–R20

27 'using aileron control': Walter Newmark, 'Ace Soarer: Observations on
the flight of the fulmar petrel', *Sailplane and Gilder*, vol. XIII, 4 May 1945,
pp. 6–7, quoted in James Fisher, *The Fulmar*, New Naturalist Monograph
No. 6, Collins, 1952, pp. 5–6; James Fisher and George Waterston, 'The

Breeding Distribution, History and Population of the Fulmar (*Fulmarus glacialis*) in the British Isles', *Journal of Animal Ecology* 10.2 (Nov. 1941): 204–72

28 Eynhallow in 2012: Ewan W. J. Edwards et al., 'Tracking a Northern Fulmar from a Scottish Nesting Site to the Charlie-Gibbs Fracture Zone: Evidence of Linkage between Coastal Breeding Seabirds and Mid-Atlantic Ridge Feeding Sites', *Deep Sea Research Part II: Topical Studies in Oceanography* 98. Part B (Dec. 2013): 438–44

30 'transport to mid-ocean': Fisher, *The Fulmar*, p. 5

35 maybe its name: D. Osborn and M. P. Harris, 'Organ Weights and Body Composition in Three Seabird Species', *Ornis Scandinavica (Scandinavian Journal of Ornithology)* 15.2 (June 1984): 95–7

35 different wind conditions: Robert W. Furness and David M. Bryant, 'Effect of Wind on Field Metabolic Rates of Breeding Northern Fulmars', *Ecology* 77.4 (June 1996): 1181–8

38 '*urg-urk-wib*': Fisher, *The Fulmar*, pp. 326–9

39 'counting themselves, dancing': C. A. Duffy, 'Girlfriends', *Collected Poems*, Picador, 2015, p. 165

40 astonishingly effective: R. A. Phillips and K. C. Hamer, 'Lipid Reserves, Fasting Capability and the Evolution of Nestling Obesity in Procellariiform Seabirds', *Proceedings: Biological Sciences* 266 (July 1999): 1329–34

41 'the said iles': Sir Donald Monro, *A Description of the Western Isles of Scotland Called Hybrides*, 1549 (1774), p. 158, online at http://www.undiscoveredscotland.co.uk/usebooks/monro-westernislands/

42 get nothing else: Màiri Sìne Chaimbeul, 'The Sea as an Emotional Landscape in Scottish Gaelic Song', *Proceedings of the Harvard Celtic Colloquium* 22 (2002): 70–1

43–4 'under the girdle': Rev. Neil Mackenzie, c. 1830s, quoted in Fisher, *The Fulmar*, pp. 136–7

44 'it were snowing': Rev. Neil Mackenzie, quoted in ibid., p. 137

45 a single stone: ibid., p. 140

45 the twentieth century: W. A. Montevecchi and A. K. Hufthammer, 'Zooarchaeological Implications for Prehistoric Distributions of Seabirds along the Norwegian Coasts', *Arctic* 43.2 (June 1990): 110–14

46 'toothache or boils': Martin Martin, *A Late Voyage to St Kilda*, 1698, Chapter II, online at http://www.undiscoveredscotland.co.uk/usebooks/martin-stkilda/

46 'sweet to his nose': Fisher, *The Fulmar*, p. 144

46 at least 60,000 years: C. Finlayson et al., 'Birds of a Feather: Neanderthal Exploitation of Raptors and Corvids', *PLoS ONE* 7.9 (2012): e45927

46 struggle of existence: Marcus Brittain and Nick Overton, 'The Significance of Others: A Prehistory of Rhythm and Interspecies Participation', *Society & Animals* 21 (2013): 134–49

47 'over the islands': John MacCulloch, *The Highlands and Western Isles of Scotland*, Longman, Hurst, 1824, vol. 2, p. 196

47 peat and turf fires: Andrew A. Meharg et al., 'Ancient Manuring Practices Pollute Arable Soils at the St Kilda World Heritage Site, Scottish North Atlantic', *Chemosphere* 64.11 (2006): 1818–28

48 'hear their music!': quoted in Roger Hutchinson, *St Kilda: A People's History*, Birlinn, 2014, p. 165

49 the open grave: John R. Baldwin, 'Sea Bird Fowling in Scotland and Faroe', *Folk Life* 12.1 (1974): 60–103; John R. Baldwin, 'A Sustainable Harvest: Working the Bird Cliffs of Scotland and the Western Faroes', in *Traditions of Sea-Bird Fowling in the North Atlantic Region*, Islands Book Trust, Isle of Lewis, 2005, pp. 114–61; John R. Baldwin, 'Harvesting Seabirds and their Eggs on the Irish Sea Islands (Part 4: Environmental and Cultural Influences)', *Folk Life* 50.1 (2012): 51–71; Aevar Petersen, 'Traditional Seabird Fowling in Iceland', in *Traditions of Sea-Bird Fowling in the North Atlantic Region*, Islands Book Trust, Isle of Lewis, 2005, pp. 194–215; Patricia Lysaght, 'Food-Provision Strategies on the Great Blasket Island: Sea-bird Fowling', in *Food From Nature: Attitudes, Strategies and Culinary Practices*, Uppsala: Kungl. Gustav Adolfs Akademien för svensk folkkultur, 2000, pp. 333–63

49 relationship now online: School of Scottish Studies SA1951.10.7, online at http://research.culturalequity.org/rc-b2/get-audio-detailed-recording.do?recordingId=7922; Stuart A. Harris-Logan, 'Nuair a bha Gaidhlig aig na h-eoin: An Investigation into the Art and Artifice of Avifaunal Mimesis as a Mode of Artistic Expression in Gaelic Oral Culture from the Seventeenth Century to the Present', MPhil(R) thesis, University of Glasgow, 2007, online at http://theses.gla.ac.uk/741/

52 the world today: H. Rink and F. Boas, 'Eskimo Tales and Songs', *Journal of American Folklore* 2.5 (April–June 1889): 123–31; Daniel Merkur, 'Breath-Soul and Wind Owner: The Many and the One in Inuit Religion', *American Indian Quarterly* 7.3 (Summer 1983): 23–39

2. Puffin

57 100 million years old: Hai-lu You et al., 'A Nearly Modern Amphibious Bird from the Early Cretaceous of Northwestern China', *Science* 312.5780 (2006): 1640–3

57 shape of the world: Alan Cooper and David Penny, 'Mass Survival of Birds across the Cretaceous-Tertiary Boundary: Molecular Evidence', *Science* 275.5303 (1997): 1109–13

57 giant serrated jaws: B. Bower, 'Fossil Skeleton Sets Seabird Size Record', *Science News* 132.20 (1987): 310

58 2.5 million years ago: G. Bartoli et al., 'Final Closure of Panama and the Onset of Northern Hemisphere Glaciation', *Earth and Planetary Science Letters* 237 (2005): 33–44; G. H. Haug and R. Tiedemann, 'Effect of the

Formation of the Isthmus of Panama on Atlantic Ocean Thermohaline Circulation', *Nature* 393 (June 1998): 673–6

59 only rarely evolved: John C. Briggs, 'A Faunal History of the North Atlantic Ocean', *Systematic Zoology* 19.1 (March 1970): 19–34

60 before it closed: Gerald Mayr and Thierry Smith, 'A Fossil Albatross from the Early Oligocene of the North Sea Basin', *The Auk* 129.1 (Jan. 2012): 87–95; G. J. Dyke et al., 'The Pliocene *Phoebastria* ("*Diomedea*") *anglica*: Lydekker's English Fossil Albatross', *Ibis* 149.3 (2007): 626–31

61 its cousin, *Pinguinus alfrednewtonii*: N. Adam Smith and Julia A. Clarke, 'Systematics and Evolution of the Pan-Alcidae (Aves, Charadriiformes)', *Journal of Avian Biology* 46.2 (March 2015): 125–40, online at http://paleosmith.org/pdf/Smith_Clarke_2014b.pdf

62 transformation of the organism: see Colin G. Scanes (ed.), *Sturkie's Avian Physiology*, 6th edn, Amsterdam: Elsevier, 2015

63 last summer: S. Rani and V. Kumar, 'Avian Circannual Systems: Persistence and Sex Differences', *General and Comparative Endocrinology* 190 (2013): 6–67; R. R. Thompson and E. Adkins-Regan, 'Photoperiod Affects the Morphology of a Sexually Dimorphic Nucleus within the Pre-optic Area of Male Japanese Quail', *Brain Research* 667.2 (1994): 201–8

63 incubate the egg: Michael P. Harris, Sarah Wanless and Jens-Kjeld Jensen, 'When are Atlantic Puffins *Fratercula arctica* in the North Sea and around the Faroe Islands Flightless?', *Bird Study* 61:2 (2014): 182–92; Michael P. Harris et al., 'The Winter Diet of the Atlantic Puffin *Fratercula arctica* around the Faroe Islands', *Ibis* 157.3 (2015): 468–79

64 to mate with: Claire Doutrelant et al., 'Colouration in Atlantic Puffins and Blacklegged Kittiwakes: Monochromatism and Links to Body Condition in Both Sexes', *Journal of Avian Biology* 44.5 (June 2013): 451–60; O. J. Torrissen and R. Christiansen, 'Requirements for Carotenoids in Fish Diets', *Journal of Applied Ichthyology* 11.3–4 (Dec. 1995): 225–30

65 to breed with: Amotz Zahavi, 'Mate Selection – A Selection for a Handicap', *Journal of Theoretical Biology* 53.1 (Sept. 1975): 205–14

66 paired-up sex: Tycho Anker-Nilssen et al., 'No Evidence of Extra-Pair Paternity in the Atlantic Puffin *Fratercula arctica*', *Ibis* 150.3 (2008): 619–22

67 on their way: for many of these topics and issues see Mike P. Harris and Sarah Wanless, *The Puffin*, illustrated by Keith Brockie, New Haven: Yale University Press, 2011

71 'the end for what?': David Crane, *Spectator*, 4 October 2014

72 need to catch: Karin Eilertsen, Robert T. Barrett and Torstein Pedersen, 'Diet, Growth and Early Survival of Atlantic Puffin (*Fratercula arctica*) Chicks in North Norway', *Waterbirds* 31.1 (March 2008): 107–14; Akiko Shoji, Annette L. Fayet and Kyle Elliott, 'Foraging Behaviour of Sympatric Razorbills and Puffins', *Marine Ecology Progress Series* 520 (2015): 257–67

76 their own survival: Alistair Dawson, Verdun M. King, George E. Bentley and Gregory F. Ball, 'Photoperiodic Control of Seasonality in Birds', *Journal of Biological Rhythms* 16 (2001): 365

76 changed all that: Tim Guilford et al., 'A Dispersive Migration in the Atlantic Puffin and its Implications for Migratory Navigation', *PLoS ONE* 6.7 (2011): e21336

3. Kittiwake

82 speaks to him: *Odyssey* V.333ff.

83 their inland sea: Medmaravis and Xaver Monbailliu, *Mediterranean Marine Avifauna: Population Studies and Conservation*, Berlin: Springer, 2013, pp. 75–7

83 'dark wave hid her': *Odyssey* V.351–3

84 darkness of their eyes: John Coulson, *The Kittiwake*, A & C Black, 2011, p. 16

85 'the Herring Gull': letter to Ernst Mayr, July 1952, quoted in Richard W. Burkhardt, *Patterns of Behavior: Konrad Lorenz, Niko Tinbergen, and the Founding of Ethology*, University of Chicago Press, 2005, p. 413

85 thousands of hours: Burkhardt, *Patterns of Behavior*, p. 414

85 published in 1957: Esther Cullen, 'Adaptations in the Kittiwake to Cliff Nesting', *Ibis* 99.2 (1957): 275–302; Animal Behavior Society, *Foundations of Animal Behavior: Classic Papers with Commentaries*, University of Chicago Press, 1996, pp. 713–40

87 nest and wife are his: from Tinbergen's manuscript notebooks (3 May 1954 MS.Eng.d.2387.B.16), quoted in Graeme Robert Beale, 'Tinbergian Practice, Themes and Variations: The Field and Laboratory Methods and Practice of the Animal Behaviour Research Group under Nikolaas Tinbergen at Oxford University', PhD thesis, University of Edinburgh, 2008, p. 67

89 'and other gulls': Niko Tinbergen, *Curious Naturalists*, Country Life, 1958, p. 203

89 'at ease once more': Niko Tinbergen, *Curious Naturalists*, revised edn, Penguin, 1974, p. 249; Heather M. C. McLannahan, 'Some Aspects of the Ontogeny of Cliff Nesting Behaviour in the Kittiwake (*Rissa tridactyla*) and the Herring Gull (*Larus argentatus*)', *Behaviour* 44.1–2 (1973): 36–88

91 'prospects of sport': *The Diary of Colonel Peter Hawker*, Longman, 1893, vol. 1, p. 91, entry for 13 January 1814

92 '1 dusky grebe': ibid., p. 92, entry for 14 January 1814

92 'people eat them': John Fenton of Chesterbank in *The Statistical Account of Scotland*, Edinburgh: W. Creech, 1794, vol. 12, p. 57

93 killing them for fun: George Muirhead, *The Birds of Berwickshire*, Edinburgh: David Douglas, 1895, vol. 2, p. 290

93 'and in anguish': Charles Waterton, 'Notes of a Visit to the Haunts of the Guillemot', in J. C. Loudon (ed.), *The Magazine of Natural History*, vol. 8, Longman, Rees, 1835, p. 165

93 'in a week': *Guardian*, 18 November 1868, quoted in Jeremy Gaskell, *Who Killed the Great Auk?*, Oxford University Press, 2000, p. 172

94 *Tribune* in 1910: 'The Feather Rage', *Los Angeles Times*, 4 January 1891; Marian Martineau, 'Present the Season of Plumes', *Chicago Daily Tribune*, 9 January 1910

94 member of the audience: Lou Taylor, Annebella Pollen and Charlotte Nicklas (eds), *Profiles of the Past: Silhouettes, Fashion and Image 1760–1960*, University of Brighton/Regency Town House, 2013, p. 18

94 bird-feather hair-do: Sweetman was a friend of Wolf Tone, the Irish patriot, and his feather-death is described with references to newspapers in *The Writings of Theobald Wolfe Tone, 1763–98*, vol. 2: *America, France and Bantry Bay, August 1795 to December 1796*, ed. T. W. Moody, R. B. McDowell and C. J. Woods, Clarendon Press, 1998, p. 98

94 Forth, as food: J. C. Coulson, 'The Status of the Kittiwake in the British Isles', *Bird Study* 10:3 (1963): 147–79

95 regular mortality: ibid.; David Evans, *A History of Nature Conservation in Britain*, Routledge, 2002, p. 35

97 'on her forehead': A. F. R. Wollaston, *Life of Alfred Newton*, New York: E. P. Dutton, 1921, pp. 137–9

98 '(Seabirds and grebes)': quoted in Robin W. Doughty, *Feather Fashions and Bird Preservation*, University of California Press, 1975, pp. 20–3

99 sixteen uninterfered with: Karl-Otto Jacobsen, Kjell Einar Erikstad and Bernt-Erik Sæther, 'An Experimental Study of the Costs of Reproduction in the Kittiwake *Rissa tridactyla*', *Ecology* 76.5 (July 1995): 1636–42

101 yet to be explained: T. H. Clutton-Brock, *The Evolution of Parental Care*, Princeton University Press, 1991

102 water are higher: David B. Irons, 'Foraging Area Fidelity of Individual Seabirds in Relation to Tidal Cycles and Flock Feeding', *Ecology* 79.2 (March 1998): 647–55

103 one-way mirror glass: Jana Kotzerka, Stefan Garthe and Scott A. Hatch, 'GPS Tracking Devices Reveal Foraging Strategies of Black-Legged Kittiwakes', *Journal für Ornithologie/Journal of Ornithology* 151.2 (2009): 459–67

104 dying in the winter: Maria I. Bogdanova, Francis Daunt, Mark Newell, Richard A. Phillips, Michael P. Harris and Sarah Wanless, 'Seasonal Interactions in the Black-Legged Kittiwake, *Rissa tridactyla*: Links between Breeding Performance and Winter Distribution', *Proceedings: Biological Sciences* 278 (Aug. 2011): 2412–18

105 miles in extent: Rachael A. Orben et al., 'Wintering North Pacific Black-Legged Kittiwakes Balance Spatial Flexibility and Consistency', *Movement Ecology* 3.1 (2015): 36

107 cataclysmic declines: for an overall picture of the decline see http://jncc. defra.gov.uk/page-2889; M. Frederiksen, M. P. Harris, F. Daunt, P. Rothery

and S. Wanless, 'The Role of Industrial Fisheries and Oceanographic Change in the Decline of North Sea Black-Legged Kittiwakes', *Journal of Applied Ecology* 41 (2004): 1129–39

108 the young fledged: K. C. Hamer et al., 'The Influence of Food Supply on the Breeding Ecology of Kittiwakes *Rissa tridactyla* in Shetland', *Ibis* 135.3 (1993): 255–63

108 50 miles away: Francis Daunt et al., 'Foraging Strategies of the Black-Legged Kittiwake *Rissa tridactyla* at a North Sea Colony: Evidence for a Maximum Foraging Range', *Marine Ecology Progress Series* 245 (2002): 239–47

108 'lack of sandeels': Martin Heubeck, 'The Decline of Shetland's Kittiwake Population', *British Birds* 95 (March 2002): 118–22

108 coast of Norway: A. Ponchon et al., 'When Things Go Wrong: Intra-Season Dynamics of Breeding Failure in a Seabird', *Ecosphere* 5.1 (2014): 4

109 will be fine: Philippe M. Cury et al., 'Global Seabird Response to Forage Fish Depletion — One-Third for the Birds', *Science* 334.6063 (2011): 1703

111 west of Alaska: Barbara M. Braun and George L. Hunt Jr, 'Brood Reduction in Black-Legged Kittiwakes', *The Auk* 100.2 (April 1983): 469–76

114 failure to breed: A. Ponchon et al., 'Tracking Prospecting Movements Involved in Breeding Habitat Selection: Insights, Pitfalls and Perspectives', *Methods in Ecology and Evolution* 4 (2013): 143–50

4. Gull

119 of new creation: Aevar Petersen, 'Formation of a Bird Community on a New Island, Surtsey, Iceland', *Surtsey Research* 12 (2009): 133–48

120 on the sand: Finnur Gudmundsson, 'Bird Observations on Surtsey in 1966', *Surtsey Research Progress Report* 3 (1967): 37–41

122 Van Gogh: first the pigeons were given food if they chose Van Gogh paintings and discouraged if they went for Chagall. When they were shown paintings they hadn't seen before, they were still able to distinguish between the two painters and went on choosing the Van Goghs. They later repeated the trick with Monet and Picasso. S. Watanabe, 'Van Gogh, Chagall and Pigeons: Picture Discrimination in Pigeons and Humans', *Animal Cognition* 4 (2001): 147–51; S. Watanabe, M. Wakita and J. Sakamoto, 'Discrimination of Monet and Picasso in Pigeons', *Journal of the Experimental Analysis of Behavior* 63 (1995): 165–74

122 'a coat pocket': R. M. Strong, 'On the Habits and Behavior of the Herring Gull, *Larus argentatus*', *The Auk* 31.1 (Jan. 1914): 22–49

124 ate the bread: Pierre-Yves Henry and Jean-Christophe Aznar, 'Tool-Use in Charadrii: Active Bait-Fishing by a Herring Gull,' *Waterbirds* 29.2 (June 2006): 233–4

125 unblinking eye: Dariusz Jakubas and Katarzyna Wojczulanis-Jakubas, 'Glaucous Gull Predation on Dovekies: Three New Hunting Methods', *Arctic* 63.4 (Dec. 2010): 468–70

127 set of answers: Niko Tinbergen, 'Comparative Studies of the Behaviour of Gulls (*Laridae*): A Progress Report', *Behaviour* 15.1 (1960): 1–69; Niko Tinbergen, 'On Adaptive Radiation in Gulls (Tribe Larini)', *Zoologische Mededelingen* 39 (1963): 209–23

127 they were dark: G. C. Phillips, 'Survival Value of the White Coloration of Gulls and Other Sea Birds', DPhil thesis, University of Oxford, 1962, quoted in Tinbergen, 'On Adaptive Radiation in Gulls (Tribe Larini)', p. 212

127 'mirror' of white: Not everyone buys this theory. Lance Tickell has pointed out that white is the default colour of feathers. The difficult question is why some birds are *not* white. WLN Tickell, 'White Plumage', *Waterbirds* 26.1 (March 2003): 1–12

129 few days older: Jasper Parsons, 'Cannibalism in Herring Gulls', *British Birds* 64 (Dec. 1971): 528–37

133 altruism in gulls: Raymond Pierotti, 'Spite and Altruism in Gulls', *American Naturalist* 115.2 (1980): 290–300

135 end of the state: Thomas S. Roberts, 'An Account of the Nesting Habits of Franklin's Rosy Gull (*Larus franklinii*), as Observed at Heron Lake in Southern Minnesota', *The Auk* 17.3 (July 1900): 272–83

137 survival strategies: P. N. Hébert, 'Adoption Behaviour by Gulls: A New Hypothesis', *Ibis* 130.2 (1988): 216–20

138 an entire population: J. W. F. Davis and E. K. Dunn, 'Intraspecific Predation and Colonial Breeding in Lesser Black-Backed Gulls *Larus fuscus*', *Ibis* 118.1 (1976): 65–7

142 a million individuals: Peter Rock, 'Urban Gulls', *British Birds* 98 (July 2005): 338–55

144 house-mites or rats: ibid.

144 at a cost: Heidi J. Auman, Catherine E. Meathrel and Alastair Richardson, 'Supersize Me: Does Anthropogenic Food Change the Body Condition of Silver Gulls? A Comparison between Urbanized and Remote, Non-Urbanized Areas', *Waterbirds* 31.1 (March 2008): 122–6

145 making junk birds: Cynthia A. Annett and Raymond Pierotti, 'Long-Term Reproductive Output in Western Gulls: Consequences of Alternate Tactics in Diet Choice', *Ecology* 80.1 (Jan. 1999): 288–97

146 unwelcome side-effects: Jerrold L. Belant, 'Gulls in Urban Environments: Landscape-Level Management to Reduce Conflict', *Landscape and Urban Planning* 38.3–4 (1997): 245–58

146 like that either: Peter Rock, 'Urban Gulls: Why Current Control Methods Always Fail', *Rivista Italiana di Ornitologia* 82.1–2 (2013) 58–65; R. A. Dolbeer et al., 'Shooting Gulls Reduces Strikes with Aircraft at John F. Kennedy International Airport', *Wildlife Society Bulletin* 21 (1993): 442–50; Thomas W. Seamans and Jerrold L. Belant, 'Comparison of DRC-1339 and Alpha-Chloralose to Reduce Herring Gull Populations', *Wildlife Society Bulletin* 27 (1999): 729–33

146 used by seaplanes: Scottish Government, *Review of Urban Gulls and their Management in Scotland*, 2006, online at http://www.gov.scot/Publications/2006/05/18113519/0; Associated Press archive, 'Three Little Island Pigs Not So Little Anymore', 2 June 1993

147 'small fry and garbage': John R. Baldwin, 'Harvesting Seabirds and their Eggs: 4', *Folk Life* 50.1 (May 2012): 60; Sir George Head, *A Home Tour through the Manufacturing Districts of England in the Summer of 1835*, John Murray, 1836

148 his kitchen door?: for the whooper swan, see Nick J. Overton and Yannis Hamilakis, 'A Manifesto for a Social Zooarchaeology: Swans and Other Beings in the Mesolithic', *Archaeological Dialogues* 20.2 (2013): 111–36; for the sea-eagle tomb, W. Hedges (ed.), *Isbister: A Chambered Tomb in Orkney*, BAR British Series 115, Oxford: British Archaeological Reports, 1983, pp. 133–50; for the protected skuas, Robert W. Furness, *The Skuas*, A & C Black, 2010, pp. 55, 283

5. Guillemot

152 of my skull: M. de L. Brooke, S. Hanley and S. B. Laughlin, 'The Scaling of Eye Size with Body Mass in Birds', *Proceedings: Biological Sciences* 266 (Feb. 1999): 405–12

153 'on the living': Hugh MacDiarmid, *Selected Poetry*, New York: New Directions, 2006, p. 178

156 'of the earth': H. Beston, *The Outermost House: A Year of Life on the Great Beach of Cape Cod*, 1928 (1988), p. 25

156 million breeding guillemots: Paul M. Regular, William A. Montevecchi and April Hedd, 'Must Marine Predators Always Follow Scaling Laws? Memory Guides the Foraging Decisions of a Pursuit-Diving Seabird', *Animal Behaviour* 86.3 (Sept. 2013): 545–52

157 with GPS trackers: Henri Weimerskirch et al., 'Use of Social Information in Seabirds: Compass Rafts Indicate the Heading of Food Patches', *PLoS ONE* 5.3 (2010): e9928

158 Witless Bay in Newfoundland: Alan E. Burger, 'Arrival and Departure Behavior of Common Murres at Colonies: Evidence for an Information Halo?', *Colonial Waterbirds* 20.1 (1997): 55–65

161 the vulnerable chicks: T. R. Birkhead and D. N. Nettleship, 'Alloparental Care in the Common Murre (*Uria aalge*)', *Canadian Journal of Zoology* 62.11 (1984): 2121–4

162 eighty-five Brünnich's guillemots: V. L. Friesen, W. A. Montevecchi, A. J. Gaston, R. T. Barrett and W. S. Davidson, 'Molecular Evidence for Kin Groups in the Absence of Large-Scale Genetic Differentiation in a Migratory Bird', *Evolution* 50.2 (April 1996): 924–30

163 looking after itself: W. D. Hamilton, 'The Genetical Evolution of Social Behaviour. II', *Journal of Theoretical Biology* 7.1 (1964): 17–52

163 on the margins: Hanna Kokko, Michael P. Harris and Sarah Wanless,
     'Competition for Breeding Sites and Site-Dependent Population Regulation
     in a Highly Colonial Seabird, the Common Guillemot *Uria aalge*', *Journal of
     Animal Ecology* 73.2 (March 2004): 367–76

164 'out at sea': Tim Birkhead, *Bird Sense: What It's Like to be a Bird*, Bloomsbury,
     2012, p. 9

165 for the stage: Allison T. Moody, Sabina I. Wilhelm, Maureen L. Cameron-
     MacMillan, Carolyn J. Walsh and Anne E. Storey, 'Divorce in Common
     Murres (*Uria aalge*): Relationship to Parental Quality', *Behavioral Ecology and
     Sociobiology* 57.3 (2005): 224–30; Carolyn J. Walsh, Sabina I. Wilhelm,
     Maureen L. Cameron-MacMillan and Anne E. Storey, 'Extra-Pair
     Copulations in Common Murres I: A Mate Attraction Strategy?', *Behaviour*
     143.10 (Oct. 2006): 1241–62; B. J. Hatchwell, 'Intraspecific Variation in
     Extra-Pair Copulation and Mate Defence in Common Guillemots *Uria
     aalge*', *Behaviour* 107.3–4 (Dec. 1988): 157–85; Tim R. Birkhead et al.,
     'Extra-Pair Paternity in the Common Murre', *The Condor* 103.1 (2001):
     158–62

169 guillemot world apart: Kate Ashbrook et al., 'Hitting the Buffers:
     Conspecific Aggression Undermines Benefits of Colonial Breeding under
     Adverse Conditions', *Biology Letters* 4.6 (2008): 630–3

171 'to the skin': Jean M. Murray (ed.), *The Newfoundland Journal of Aaron
     Thomas, Able Seaman in HMS 'Boston': A Journal Written during a Voyage from
     England to Newfoundland*, Longman, 1968, p. 1

171 the Red Indians: Ingeborg Marshall, *A History and Ethnography of the Beothuk*,
     McGill-Queen's University Press, 1998, p. 113

172 'crowne of the head': ibid., p. 30

172 an integrated universe: Todd J. Kristensen, 'Seasonal Bird Exploitation by
     Recent Indian and Beothuk Hunter-Gatherers of Newfoundland', *Canadian
     Journal of Archaeology / Journal Canadien d'Archéologie* 35.2 (2011): 292–322

173 into winter food: Marshall, *A History and Ethnography of the Beothuk*, pp. 67,
     295

173 weakened by malnutrition: Donald H. Holly Jr, 'The Beothuk on the Eve of
     their Extinction', *Arctic Anthropology* 37.1 (2000): 79–95; Marshall, *A History
     and Ethnography of the Beothuk*, p. 137

173 'an island afterlife': Todd J. Kristensen and Donald H. Holly Jr, 'Birds,
     Burials and Sacred Cosmology of the Indigenous Beothuk of Newfoundland,
     Canada', *Cambridge Archaeological Journal* 23.1 (2013): 41–53

174 deer-sinew stitches: J. P. Howley, *The Beothucks or Red Indians*, Cambridge
     University Press, 1915, pp. 331–2

175 the world below: Kristensen and Holly, 'Birds, Burials and Sacred
     Cosmology of the Indigenous Beothuk of Newfoundland, Canada',
     pp. 41–53

176 'their island afterworld': ibid., p. 50

176　the Yup'ik: Ann Fienup-Riordan, 'The Bird and the Bladder: The Cosmology of Central Yup'ik Seal Hunting', *Etudes/Inuit/Studies* 14.1–2 (1990): 23–38

177　ice-scattered sea: Ann Fienup-Riordan, 'Metaphors of Conversion, Metaphors of Change', *Arctic Anthropology* 34.1 (1997): 102–16

6. Cormorant and Shag

179　'he fixes sad': *Paradise Lost* IV.28; 'my self am Hell': IV.75; 'his tumultuous breast': IV.16

179　'damaskt with flours': *Paradise Lost* IV.334

180　'old Ocean smiles': *Paradise Lost* IV.159–65

180　'with Golden Rinde': *Paradise Lost* IV.248–9

180　'eminent, blooming Ambrosial': *Paradise Lost* IV.218–19; 'Nectarine Fruits': IV.332

180　'them who liv'd': *Paradise Lost* IV.158–9

181　what others had: see Richard J. King, *The Devil's Cormorant*, University of New Hampshire Press, 2013

181　'Shylock': the King James Bible translates the word רָלָשָׁה, as 'the cormorant'. 'Shaloch' transliterates that Hebrew word. See Sir Israel Gollancz, 'Man is What Man Had Made Him', 1916, in William Baker and Brian Vickers (eds), *Shakespeare: The Critical Tradition: The Merchant of Venice*, Bloomsbury, 2005, p. 298

183　'flying cheerfully away!': W. S. M. D'Urban and the Rev. Murray A. Mathew, *The Birds of Devon*, R. H. Pouter, 1892, p. 176

184　from their partner: B. W. Tucker, 'Brood-Patches and the Physiology of Incubation', *British Birds* 37 (July 1943): 22–8

186　'abundant digestion': Pierre-Amédée Pichot, *Les Oiseaux de sport*, Paris: Librairie Ad. Legoupy, 1903, p. 27

187　fish back unasked: Berthold Laufer, 'The Domestication of the Cormorant in China and Japan', *Publications of the Field Museum of Natural History. Anthropological Series* 18.3 (1931): 201–62; Sir George Staunton, *An Authentic Account of an Embassy from the King of Great Britain to the Emperor of China*, G. Nicol, 1797, vol. 2, p. 388

187　'not objective characteristics': G. E. Freeman and F. H. Salvin, *Falconry: Its Claims, History, and Practice. To which are Added Remarks on Training the Otter and Cormorant*, Longman, Green, 1859, p. 331

188　where they fished: Laufer, 'The Domestication of the Cormorant in China and Japan', p. 207

188　'how smart animals are?': Frans de Waal, *Are We Smart Enough to Know How Smart Animals Are?*, London and New York: W. W. Norton, 2016

189　are temporarily away: G. R. Potts, J. C. Coulson and I. R. Deans, 'Population Dynamics and Breeding Success of the Shag, *Phalacrocorax aristotelis*, on the Farne Islands, Northumbria', *Journal of Animal Ecology* 49.2 (June 1980): 465–84

191 the European shag: Frédérique Dubois, Frank Cézilly and Mark Pagel, 'Mate Fidelity and Coloniality in Waterbirds: A Comparative Analysis', *Oecologia* 116.3 (1998): 433–40

191 size of the brain: N. J. Emery and N. Clayton, 'The Mentality of Crows: Convergent Evolution of Intelligence in Corvids and Apes', *Science* 306.5703 (2004): 1903–7

191 a bird's lifestyle: Nathan J. Emery, Amanda M. Seed, Auguste M. P. von Bayern and Nicola S. Clayton, 'Cognitive Adaptations of Social Bonding in Birds', *Philosophical Transactions: Biological Sciences* 362 (May 2007): 489–505

193 understand its world: Robert E. Ricklefs, 'The Cognitive Face of Avian Life Histories: The 2003 Margaret Morse Nice Lecture', *Wilson Bulletin* 116.2 (June 2004): 119–33

193 'for both parties': Emery et al., 'Cognitive Adaptations of Social Bonding in Birds', p. 502

194 understanding and sociability: Seweryn Olkowicz et al., 'Birds Have Primate-Like Numbers of Neurons in the Forebrain', *Proceedings of the National Academy of Sciences* 113.26 (June 2016): 7255–60

195 habits just offshore: Gal Ribak, Daniel Weihs and Zeev Arad, 'Water Retention in the Plumage of Diving Great Cormorants *Phalacrocorax carbo sinensis*', *Journal of Avian Biology* 36.2 (March 2005): 89–95

195 buoyancy during dives: R. Stephenson, D. L. Turner and P. J. Butler, 'The Relationship between Diving Activity and Oxygen Storage Capacity in the Tufted Duck (*Aythya fuligula*)', *Journal of Experimental Biology* 141 (1989): 265–75

196 remain almost constant: A. M. Rijke, 'The Water Repellency and Feather Structure of Cormorants', *Journal of Experimental Biology* 48 (1968): 185–9; David Grémillet et al., 'Body Temperature and Insulation in Diving Great Cormorants and European Shags', *Functional Ecology* 12 (1998): 386–94; David Grémillet et al., 'Unusual Feather Structure Allows Partial Plumage Wettability in Diving Great Cormorants *Phalacrocorax carbo*', *Journal of Avian Biology* 36.1 (Jan. 2005): 57–63

196 successor and replacement: figures at http://www.garden-birds.co.uk/information/lifespan.htm

197 coast of Scotland: F. Daunt et al., 'From Cradle to Early Grave: Juvenile Mortality in European Shags *Phalacrocorax aristotelis* Results from Inadequate Development of Foraging Proficiency', *Biology Letters* 3.4 (2007): 371–4

199 even as adults: M. Frederiksen et al., 'The Demographic Impact of Extreme Events: Stochastic Weather Drives Survival and Population Dynamics in a Long-Lived Seabird', *Journal of Animal Ecology* 77.5 (Sept. 2008): 1020–91

7. Shearwater

203 'barren salt sea': *Odyssey* V.69–74 (author's own translation)

203 'concave towering high': *Paradise Lost* II.629–35

205 coast in 1936: for the following pages see R. M. Lockley, *Shearwaters*,
     J. M. Dent, 1942

205 'the constant freeholders': ibid., p. xi

206 'has fairly begun': ibid., p. 5

206 Skokholm every year: on inadequacy of previous books see ibid., p. 10

206 nibbling from below: ibid., pp. 117–33

207 Caroline had laid: ibid., pp. 147ff

208 on further experiments: ibid., p. 154

208 'their homing spurs': Michael Brooke, *The Manx Shearwater*, T & AD Poyser,
     2010, p. 66

209 from the east: for Venice experiments see Lockley, *Shearwaters*,
     pp. 174–87

209 'the long voyage': ibid., p. 176

210 for its adventures: David Lack and R. M. Lockley, 'Skokholm Bird
     Observatory Homing Experiments', *British Birds* 31 (Jan. 1938): 242–8

211 'survived the journey': Rosario Mazzeo, 'Homing of the Manx Shearwater',
     The Auk 70.2 (April 1953), p. 200

211 'beating the mail!': ibid., p. 201

212 'true, bi-co-ordinate navigation': G. V. T. Matthews, 'Navigation in the Manx
     Shearwater', *Journal of Experimental Biology* 30 (1953): 370–96

214 'and general excitement': ibid., p. 382

215 properly guessed at: C. C. Dixon suggested in 1932 that albatrosses follow
     prevailing winds around the Southern Ocean. This prompted V. C. Wynne-
     Edwards in 1935 to speculate that great shearwaters follow circular
     migration routes around the Atlantic to take advantage of prevailing winds.
     By the 1950s Bill Bourne and Nagahisa Kuroda were discussing trans-
     equatorial figure-of-eight migrations of various shearwater species. See: C.
     C. Dixon, 'Some observations on the albatrosses and other birds of the
     southern oceans', *Transactions of the Royal Canadian Institute*, 1932, 19:117–
     139; N. Kuroda, 'A brief note on the pelagic migration of the Tubinares',
     Miscellaneous Reports of the Yamashina Institute for Ornithology and Zoology,
     1957, 1:436–449; V. C. Wynne-Edwards, 'On the habits and distribution of
     birds on the North Atlantic', *Proceedings of the Boston Society of Natural
     History*, 1935, 40

216 vast yearly odyssey: Brooke, *The Manx Shearwater*, pp. 54–6. Brooke analysed
     all the rings recovered up to the end of 1987

216 was clear enough: T. Guilford et al., 'Migration and Stopover in a Small
     Pelagic Seabird, the Manx Shearwater *Puffinus puffinus*: Insights from
     Machine Learning', *Proceedings: Biological Sciences* 276 (Feb. 2009):
     1215–23

218 workings of the wind: Ángel M. Felicísimo, Jesús Muñoz and Jacob
     González-Solís, 'Ocean Surface Winds Drive Dynamics of Transoceanic
     Aerial Movements', *PLoS ONE* 3.8 (2008): e2928

218 in the north: Scott A. Shaffer et al., 'Migratory Shearwaters Integrate
Oceanic Resources across the Pacific Ocean in an Endless Summer',
Proceedings of the National Academy of Sciences 103.34 (Aug. 2006): 12799–802

219 decision-making individuals: Maria P. Dias et al., 'Breaking the Routine:
Individual Cory's Shearwaters Shift Winter Destinations between
Hemispheres and across Ocean Basins', *Proceedings: Biological Sciences* 278
(June 2011): 1786–93

219 'acres of volcanic rock': Lockley, *Shearwaters*, p. 221; 'splendid hullabaloo':
ibid., p. 222

221 their giant voyages: Bruno Massa et al., 'Homing of Cory's Shearwaters
(*Calonectris diomedea*) Carrying Magnets', *Italian Journal of Zoology* 58.3
(1991): 245–7

221 'transform and create': François Jacob, *The Possible and the Actual*, University
of Washington Press, 1982, quoted in *New Scientist*, 5 August 1982, p. 378

222 lyre-bird or cockatoo: Julie C. Hagelin and Ian L. Jones, 'Bird Odors and
Other Chemical Substances: A Defense Mechanism or Overlooked Mode of
Intraspecific Communication?', *The Auk* 124.3 (July 2007): 741–61

222 'on another cruise': Nancy Averett, 'The Sniff Test', *Audubon* 116.1 (Jan./
Feb. 2014): 46

223 'scientists live for': ibid.

224 ate the krill: Gabrielle A. Nevitt and Francesco Bonadonna, 'Seeing the
World through the Nose of a Bird: New Developments in the Sensory
Ecology of Procellariiform Seabirds', *Marine Ecology Progress Series* 287
(2005): 292–5

224 smells they know: R. W. Van Buskirk and G. A. Nevitt. 'The Influence of
Developmental Environment on the Evolution of Olfactory Foraging
Behaviour in Procellariiform Seabirds', *Journal of Evolutionary Biology* 21.1
(2008): 67–76

225 sense of seabirds: Lucia F. Jacobs, 'From Chemotaxis to the Cognitive Map:
The Function of Olfaction', *Proceedings of the National Academy of Sciences*
109.Supplement 1 (June 2012): 10693–700

225 'over evolutionary time': Cornelia I. Bargmann, 'Comparative
Chemosensation from Receptors to Ecology', *Nature* 444 (Nov. 2006):
295–301

226 only their noses: Lucia F. Jacobs et al., 'Olfactory Orientation and
Navigation in Humans', *PLoS ONE* 10.6 (2015): e0129387

227 in September 1957: Anna Gagliardo et al., 'Oceanic Navigation in Cory's
Shearwaters: Evidence for a Crucial Role of Olfactory Cues for Homing
after Displacement', *Journal of Experimental Biology* 216 (2013): 2798–805

231 mistake for food: Matthew S. Savoca, Martha E. Wohlfeil, Susan E. Ebeler
and Gabrielle A. Nevitt, 'Marine Plastic Debris Emits a Keystone
Infochemical for Olfactory Foraging Seabirds', *Science Advances* 2.11 (Nov.
2016): e1600395

8. Gannet

238 white of its wings: http://datazone.birdlife.org/species/factsheet/
Nazca-Booby

238 Galápagos since 1985: D. J. Anderson et al., 'The Role of Hatching
Asynchrony in Siblicidal Brood Reduction of Two Booby Species', *Behavioral
Ecology and Sociobiology* 25.5 (1989): 363–8; David J. Anderson et al.,
'Evolution of Obligate Siblicide in Boobies. 1. A Test of the Insurance-Egg
Hypothesis', *American Naturalist* 135.3 (1990): 334–50; Elisa M. Tarlow,
Martin Wikelski and David J. Anderson, 'Hormonal Correlates of Siblicide
in Galápagos Nazca Boobies', *Hormones and Behavior* 40.1 (2001): 14–20;
David J. Anderson, Elaine T. Porter and Elise D. Ferree, 'Non-Breeding
Nazca Boobies (*Sula granti*) Show Social and Sexual Interest in Chicks:
Behavioural and Ecological Aspects', *Behaviour* 141.8 (2004): 959–77

241 the human world: Jacquelyn K. Grace et al., 'Hormonal Effects of
Maltreatment in Nazca Booby Nestlings: Implications for the "Cycle of
Violence"', *Hormones and Behavior* 60.1 (2011): 78–85; Martina S. Müller et
al., 'Maltreated Nestlings Exhibit Correlated Maltreatment as Adults:
Evidence of a "Cycle of Violence" in Nazca Boobies (*Sula granti*)', *The Auk*
128.4 (Oct. 2011): 615–19

243 called Ailsa Craig: In Seamus Heaney, *Sweeney Astray: A Version from the Irish*,
Macmillan, 1985, 'Sweeney's Lament on Ailsa Craig', line 52

244 were talking to: William Harvey, *Exercitationes de Generatione Animalium*, Du
Gard for O. Pulleyn, 1651, p. 30

244 dark-winged savage gull: from the Browne manuscripts in the British
Library: BL MS Sloane 1830 folio 6r and BL MS Sloane 1848 folio 225v,
kindly provided by Claire Williams and Claire Preston

246 out of every four: Bryan Nelson, *The Atlantic Gannet*, 2nd edn, Great
Yarmouth: Fenix Books, 2001, pp. 134–66

247 and a debrief: Bryan Nelson, *Living with Seabirds*, Edinburgh University
Press, 1986, pp. 6–8

247 'has been punctured': ibid., pp. 42–3

248 two or three days: ibid., pp. 52–3

248 disturbing the birds: R. T. Barret, 'Recent establishment and extinctions of
Northern Gannet (*morus bassanus*) colonies in North Norway, 1995–2008,
Ornis Norvegica, 2008, 31: 172–82

249 your chicks thrive: Nelson, *The Atlantic Gannet*, pp. 103–21

249 love and affection: ibid., pp. 129–31

251 where they were fishing: Ewan D. Wakefield et al., 'Space Partitioning
without Territoriality in Gannets', *Science* 341.6141 (2013): 68–70

255 make it worth it: Etienne Danchin et al., 'Public Information: From Nosy
Neighbors to Cultural Evolution', *Science* 305.5683 (2004): 487–91;
Blandine Doligez, Etienne Danchin and Jean Clobert, 'Public Information
and Breeding Habitat Selection in a Wild Bird Population', *Science* 297.5584

(2002): 1168–70; Thomas J. Valone, 'From Eavesdropping on Performance to Copying the Behavior of Others: A Review of Public Information Use', *Behavioral Ecology and Sociobiology* 62.1 (2007): 1–14; Henri Weimerskirch, 'Seabirds – Individuals in Colonies', *Science* 341.6141 (2013): 35–6

257 life at sea?: Cecile Rolland, Etienne Danchin and Michelle de Fraipont, 'The Evolution of Coloniality in Birds in Relation to Food, Habitat, Predation, and Life-History Traits: A Comparative Analysis', *American Naturalist* 151.6 (1998): 514–29

258 ever been made: Norman MacCaig, 'Rhu Mor', in Tim Dee and Simon Armitage (eds), *The Poetry of Birds*, Penguin, 2009, p. 9

258 as a result: Nelson, *The Atlantic Gannet*, p. 295

259 wings must fold: Gabriel E. Machovsky-Capuska et al., 'Visual Accommodation and Active Pursuit of Prey Underwater in a Plunge-Diving Bird: The Australasian Gannet', *Proceedings: Biological Sciences* 279 (Aug. 2012): 4118–25

259 miracles of seabird life: André Rochon-Duvigneaud, *Les Yeux et la vision des vertébrés*, Paris: Masson, 1943, quoted in Graham Martin, *Birds by Night*, A & C Black, 2010, p. 5

259 in the shoal: Machovsky-Capuska et al., 'Visual Accommodation and Active Pursuit of Prey Underwater in a Plunge-Diving Bird: The Australasian Gannet', pp. 4118–25

261–2 'Like right an' wrang': from Katrina Porteous, *The Lost Music*, Newcastle upon Tyne: Bloodaxe, 1996, pp. 62–3

262 with GPS loggers: Samantha C. Patrick et al., 'Individual Differences in Searching Behaviour and Spatial Foraging Consistency in a Central Place Marine Predator', *Oikos* 123.1 (2014): 33–40

262 in the Celtic Sea: ibid.; Thomas W. Bodey et al., 'Seabird Movement Reveals the Ecological Footprint of Fishing Vessels', *Current Biology* 24.11 (2014): R514–R515

263 'such personalities': E. D. Wakefield et al., 'Long-term individual foraging site fidelity', Ecology 96 (2015): 3058–74

9. Great Auk and its Cousin Razorbill

270 for each pair: Clare S. Lloyd, 'Factors Affecting Breeding of Razorbills *Alca torda* on Skokholm', *Ibis* 121.2 (1979): 165–76. The combination of factors means that only 70 per cent of eggs laid produced a fledged chick

270 can be no life: Jennifer L. Lavers, Ian L. Jones and Antony W. Diamond, 'Age at First Return and Breeding of Razorbills (*Alca torda*) on the Gannet Islands, Labrador and Machias Seal Island, New Brunswick', *Waterbirds* 31.1 (March 2008): 30–4. Of more than 6,200 razorbill chicks ringed between 1995 and 2006, a total of 969 were resighted where they were ringed, a 16 per cent return. A proportion will have gone to breed at other colonies

270 'of its inhabitants': quoted in Andrew Norman, *Charles Darwin: Destroyer of Myths*, Pen and Sword, 2013, p. 37

270 'places of her childhood': Melanie Challenger, *On Extinction*, Berkeley: Counterpoint, 2012, p. 43; 'lodging in the mind': ibid., p. 55

271 'past and present': ibid., pp. 55–6

271 the great auk: for names see Jeremy Gaskell, *Who Killed the Great Auk?*, Oxford University Press, 2000, p. 7

273 'was quite dead': reported to J. A. Harvie-Brown in *Vertebrate Fauna of the Outer Hebrides*, quoted in ibid., pp. 142–3

273 the giant chick: Stewart Brand on the great auk and de-extinction, http://blog.longnow.org/02016/02/04/is-the-great-auk-a-candidate-for-de-extinction/

273 'is the exception': quoted by Eizabeth Kolbert, http://www.newyorker. com/magazine/2009/05/25/the-sixth-extinction

273 about AD 1300: see David Steadman, *Extinction and Biogeography of Tropical Pacific Birds*, University of Chicago Press, 2006

273 what they represent: see E. O. Wilson, *The Diversity of Life*, Penguin, 2001

274 from the Atlantic world: J. C. Rando, 'New Data of Fossil Birds from El Hierro (Canary Islands): Probable Causes of Extinction and Some Biogeographical Considerations', *Ardeola* 49.1 (2002): 39–49

275 Atlantic bird life: James A. Tuck, 'An Archaic Cemetery at Port au Choix, Newfoundland', *American Antiquity* 36.3 (1971): 343–58

275 three great auks: Jean Clottes et al., 'La Grotte Cosquer (Cap Morgiou, Marseille)', *Bulletin de la Société Préhistorique Française* 89.4 (1992): 98–128

276 'a man could walk': Sven-Axel Bengtson, 'The Great Auk (*Pinguinus impennis*): Anecdotal Evidence and Conjectures', *The Auk* 101.1 (Jan. 1984): 1–12

276 'the birds' size': Francesco d'Errico, 'Birds of the Grotte Cosquer: The Great Auk and Palaeolithic Prehistory', *Antiquity* 68.258 (1994): 39–47

278 his *vinho verde?*: John R. Stewart, 'Sea-birds from Coastal and Non-Coastal, Archaeological and "Natural" Pleistocene Deposits or Not All Unexpected Deposition is of Human Origin', *Acta Zoologica Cracoviensia* (Kraków), 45 (special issue) (2002): 167–78; four great auks from the Roman period have been found in the Netherlands, probably washed up on the beaches of the North Sea but possibly traded or brought back there by Batavian soldiers stationed in Britain on Hadrian's Wall. M. Groot, 'The Great Auk (*Pinguinus impennis*) in the Netherlands during the Roman Period', *International Journal of Osteoarchaeology* 15 (2005): 15–22

279 untouched nature looked like: W. Jeffrey Bolster, *The Mortal Sea*, Cambridge: Belknap Press, 2012, p. 47

280 'a paved street': ibid., p. 39

280 markets every year: Daniel Vickers (ed.), *A Companion to Colonial America*,
     Hoboken, NJ: John Wiley, 2008, p. 494

281 new colonists reported: quoted in Bolster, *The Mortal Sea*, p. 83

281 'obtained at once': Joseph William Collins, *Notes on the Habits and Methods of
     Capture of Various Species of Sea Birds that Occur on the Fishing Banks Off the
     Eastern Coast of North America*, Washington, DC: US Government Printing
     Office, 1884, quoted in Farley Mowat, *Sea of Slaughter*, Vancouver: Douglas
     & McIntyre, 2004, reprint edn 2013, p. 48

282 'to procure provisions': Bolster, *The Mortal Sea*, p. 84

282 'abound in Maine': ibid., pp. 85–6

282 'every trophic level': H. K. Lotze and I. Milewski, 'Two Centuries of
     Multiple Human Impacts and Successive Changes in a North Atlantic Food
     Web', *Ecological Applications* 14.5 (2004): 1428–47

283 'on good paper': Jean M. Murray (ed.), *The Newfoundland Journal of Aaron
     Thomas, Able Seaman in HMS 'Boston': A Journal Written during a Voyage from
     England to Newfoundland*, Longman, 1968, p. 1

283 the governor's ball: Sarah Glassford, 'Seaman, Sightseer, Storyteller, and
     Sage: Aaron Thomas's 1794 *History of Newfoundland*', *Newfoundland and
     Labrador Studies*, Jan. 2001.

283 a chimney fire: or was this a joke?

283 pain and fear: Murray (ed.), *The Newfoundland Journal of Aaron Thomas, Able
     Seaman in HMS 'Boston'*, p. 67

283 'you had cleared': quoted in Gaskell, *Who Killed the Great Auk?*, p. 100

284 'on the island': ibid., p. 113

284 of the process: Bengtson, 'The Great Auk (*Pinguinus impennis*): Anecdotal
     Evidence and Conjectures', pp. 1–12

285 'annihilation – as inspiration': D. Lutz, 'The Elegy as Shrine: Tennyson and
     "In Memoriam"', in *Relics of Death in Victorian Literature and Culture*,
     Cambridge University Press, 2015, pp. 124–5

285 'but a cry': Alfred, Lord Tennyson, *In Memoriam A.H.H. OBIIT MDCCCXXXIII*:
     54, online at www.poetryfoundation.org/poems-and-poets/poems/
     detail/45341

286 'loved at all': ibid.: 27, online at https://www.poetryfoundation.org/
     poems-and-poets/poems/detail/45336

286 John Wolley: Tim R. Birkhead and Peter T. Gallivan, 'Alfred Newton's
     Contribution to Ornithology: A Conservative Quest for Facts Rather than
     Grand Theories', *Ibis* 154.4 (2012): 887–905. The Newton MSS are in
     Cambridge University Library: see Add.9839/25/ for Wolley
     correspondence and Add.9839/2/ for GREAT AUK papers.
     Add.9839/30/59 is a printed memoir of Wolley. A. F. R. Wollaston, *Life of
     Alfred Newton*, New York: E. P. Dutton, 1921, pp. 12–13 for Wolley's
     boyhood; pp. 27–41 for his and Newton's trip to Iceland

287 'a railway truck': Wollaston, *Life of Alfred Newton*, p. 70

287 frame of mind: the notes and papers for *Ootheca Wolleyana* are in Cambridge
    University Library under Add.9839/2/GREAT AUK, including Henrik
    Grönvold's wonderful watercolours of the great auk eggs at
    Add.9839/2/281–8 and others at Add.9839/2/150–70

288 Twombly-like scribbling: these photographs in Cambridge University
    Library are under Add.9839/2/166–244. The photographs are reproduced
    in Errol Fuller's magnificent *The Great Auk*, Southborough, Kent (privately
    published), 1999

289 where it still is: Fuller, *The Great Auk*, pp. 245–7

290 poignancy of extinction: see ibid.

292 life to come: Bengtson, 'The Great Auk (*Pinguinus impennis*): Anecdotal
    Evidence and Conjectures', pp. 1–12

292 seals and sea lions: Bradley C. Livezey, 'Morphometrics of Flightlessness in
    the Alcidae', *The Auk* 105.4 (Oct. 1988): 681–98

293 no longer possible: Kyle H. Elliott et al., 'High Flight Costs, But Low Dive
    Costs, in Auks Support the Biomechanical Hypothesis for Flightlessness in
    Penguins', *Proceedings of the National Academy of Sciences* 110.23 (June 2013):
    9380–4

10. Albatross

297 'prodigy of plumage': Herman Melville, *Moby Dick*, 1851, Chapter xlii, 'The
    Whiteness of the Whale', p. 194, online at http://etcweb.princeton.edu/
    batke/moby/

298 the Southern Ocean: http://wwf.panda.org/what_we_do/endangered_
    species/albatross/

300 'farre from it': Peter Mundy, *The Travels of Peter Mundy in Europe and Asia,
    1608–1667*, Hakluyt Society, Second Series 17 (1967), vol. 5, p. 35, quoted
    in G. Barwell, 'Uses of the Albatross: Threatened Species and Sustainability',
    in Heather Yeatman (ed.), *The SInet 2010 eBook* (Proceedings of the SInet
    2009 Conference), Social Innovation Network (SInet), University of
    Wollongong, Wollongong, 2010, p. 204

301 in the squad: R. C. Murphy, *Oceanic Birds of South America*, New York:
    Macmillan, 1936, vol. 1, p. 499

301 autumn of 1797: see John Livingstone Lowes, *The Road to Xanadu: A Study in
    the Ways of the Imagination*, Boston and New York: Houghton Mifflin, 1927,
    pp. 203–14

302 'that can be conceiv'd': George Shelvocke, *A Voyage Round the World by Way of
    the Great South Sea*, J. Senex, 1726, p. 67; 'of firm land': p. 69; 'Islands of
    ice': p. 71

302 'fair wind after it': ibid., pp. 72–3

304 'laws of hospitality': S. T. Coleridge, 'Argument' of *The Rime of the Ancient
    Mariner* in *Lyrical Ballads, with a Few Other Poems*, enlarged edn, 2 vols,
    Longman, Rees, 1800

304 the nineteenth century: Graham Barwell, 'Coleridge's Albatross and the Impulse to Seabird Conservation', *Kunapipi* 29.2 (2007): 22–61, online at ro.uow.edu.au/kunapipi/vol29/iss2/5

304 kind of fleece: Barwell, 'Uses of the Albatross: Threatened Species and Sustainability', pp. 203–15

306 'fore-topgallant mast': Murphy, *Oceanic Birds of South America*, vol. 1, pp. 499–500

306 'like beavertail soup': ibid., p. 556

306 warmth and connection: ibid., pp. 544–6

307 it was shot: ibid., p. 546

307 archipelago to another: Barwell, 'Uses of the Albatross: Threatened Species and Sustainability', pp. 203–15

308 'he cannot stir': Charles Baudelaire, *The Complete Verse*, ed. and trans. Francis Scarfe, Anvil Press Poetry, 1986, vol. 1, p. 59

309 radio-tracking technologies: Etienne Benson, 'The Wired Wilderness: Electronic Surveillance and Environmental Values in Wildlife Biology', PhD thesis, Massachusetts Institute of Technology, 2008

309 'spring of 1962': Dwain W. Warner, 'Space Tracks: Bioelectronics Extends its Frontiers', *Natural History* 62 (1963): 8–15 at 12

310 birds all failed: Benson, 'The Wired Wilderness: Electronic Surveillance and Environmental Values in Wildlife Biology'

310 PhD at Montpellier: Henri Weimerskirch, 'Ecologie Comparée des Albatros des Terres Australes Françaises', PhD thesis, Académie de Montpellier, Université des Sciences et Techniques du Languedoc, 1986

310 the Southern Ocean: http://www.scilogs.fr/ecologie-science-societe/2015/04/03/la-richesse-faunistique-des-terres-australes-et-antarctiques-francaises/

311 were killing them: Henri Weimerskirch and Pierre Jouventin, 'Population Dynamics of the Wandering Albatross, *Diomedea exulans*, of the Crozet Islands: Causes and Consequences of the Population Decline', *Oikos* 49.3 (1987): 315–22

311 report their locations: Henri Weimerskirch, John Alexander Bartle, Pierre Jouventin and Jean Claude Stahl, 'Foraging Ranges and Partitioning of Feeding Zones in Three Species of Southern Albatrosses', *The Condor* 90.1 (1988): 214–19

312 covered by feathers: Henri Weimerskirch, Thierry Mougey and Xavier Hindermeyer, 'Foraging and Provisioning Strategies of Black-Browed Albatrosses in Relation to the Requirements of the Chick: Natural Variation and Experimental Study', *Behavioral Ecology and Sociobiology* 8.6 (1997): 635–43

312 to pull off: Pierre Jouventin and Henri Weimerskirch, 'Satellite Tracking of Wandering Albatrosses', *Nature* 343 (Feb. 1990): 746–8

314 'Mrs Gibson': Karin Reinke et al., 'Understanding the Flight Movements of
     a Non-Breeding Wandering Albatross, *Diomedea exulans gibsoni*, Using a
     Geographic Information System', *Australian Journal of Zoology* 46 (1998):
     171–81

317 'the mariner's hollo!': Coleridge, *The Rime of the Ancient Mariner*, Part I,
     41–4; 71–4

318 'breeze to blow!': ibid., Part II, 9–14

319 to the south: Henri Weimerskirch, Marc Salamolard, François Sarrazin and
     Pierre Jouventin, 'Foraging Strategy of Wandering Albatrosses through the
     Breeding Season: A Study Using Satellite Telemetry', *The Auk* 110.2 (April
     1993): 325–42

319 extraordinary foraging journeys: Henri Weimerskirch, C. Patrick Doncaster
     and Franck Cuenot-Chaillet, 'Pelagic Seabirds and the Marine Environment:
     Foraging Patterns of Wandering Albatrosses in Relation to Prey Availability
     and Distribution', *Proceedings: Biological Sciences* 255 (Feb. 1994): 91–7

319 of the bird: H. Weimerskirch, T. Guionnet, J. Martin, S. A. Shaffer and
     D. P. Costa, 'Fast and Fuel Efficient? Optimal Use of Wind by Flying
     Albatrosses', *Proceedings: Biological Sciences* 267 (Sept. 2000): 1869–74

323 enough to breed: H. Weimerskirch et al., 'Lifetime Foraging Patterns of the
     Wandering Albatross: Life on the Move!', *Journal of Experimental Marine
     Biology and Ecology* 450 (2014): 68–78

323 five years old: Susan Scott, S. R. Duncan and C. J. Duncan, 'Infant Mortality
     and Famine: A Study in Historical Epidemiology in Northern England',
     Journal of Epidemiology and Community Health 49 (1995): 245–52

324 warms around it?: Henri Weimerskirch, 'Changes in Wind Pattern Alter
     Albatross Distribution and Life-History Traits', *Science* 335.6065 (2012):
     211–14

325 to their deaths: for the destruction of seabirds by longlines see Orea
     R. J. Anderson et al., 'Global Seabird Bycatch in Longline Fisheries',
     Endangered Species Research 14.2 (2011): 91–106

326 in its area: Julien Collet, Samantha C. Patrick and Henri Weimerskirch,
     'Albatrosses Redirect Flight towards Vessels at the Limit of their Visual
     Range', *Marine Ecology Progress Series* 526 (2015): 199–205

327 their own salvation: Susan M. Waugh of the Museum of New Zealand on the
     new loggers at http://blog.tepapa.govt.nz/2013/05/16/solander-
     kingdom-of-the-birds/. The loggers work by detecting the signal of the
     radar emitted by vessels. They can only tell if a vessel is within 20 km of
     where the bird is, not how close they are. To get distance and bearing you
     would need two birds with loggers simultaneously, or to have satellite data
     from the vessels to calibrate with the GPS-x

327 albatrosses on Kerguelen: Samantha C. Patrick and Henri Weimerskirch,
     'Personality, Foraging and Fitness Consequences in a Long Lived Seabird',
     PLoS ONE 9.2 (2014): e87269

331 reaction is awe: ibid.; Samantha C. Patrick, Anne Charmantier and Henri Weimerskirch, 'Differences in Boldness are Repeatable and Heritable in a Long-Lived Marine Predator', *Ecology and Evolution* 3.13 (2013): 4291–9

11.The Seabird's Cry

334 occupied since 1350: Kurt Burnham et al., 'Gyrfalcon *Falco rusticolus* Post-Glacial Colonization and Extreme Long-Term Use of Nest-Sites in Greenland', *Ibis* 151.3 (2009): 514–22; Kurt Burnham and Ian Newton, 'Seasonal Movements of Gyrfalcons *Falco rusticolus* Include Extensive Periods at Sea', *Ibis* 153.3 (2011): 468–84

334 34,000 years ago: Steven D. Emslie et al., 'A 45,000 Yr Record of Adélie Penguins and Climate Change in the Ross Sea, Antarctica', *Geology* 35.1 (Jan. 2007): 61–4; Sven Thatje et al., 'Life Hung by a Thread: Endurance of Antarctic Fauna in Glacial Periods', *Ecology* 89.3 (March 2008): 682–92; A. J. Gaston and G. Donaldson, 'Peat Deposits and Thick-Billed Murre Colonies in Hudson Strait and Northern Hudson Bay: Clues to Post-Glacial Colonization of the Area by Seabirds', *Arctic* 48.4 (Dec. 1995): 354–8

335 in the 1960s: for long-term trends of British seabirds, see Joint Nature Conservation Committee, *Seabird Population Trends and Causes of Change: 1986–2015 Report*, 2016, online at http://jncc.defra.gov.uk/page-3201

336 in six decades: Michelle Paleczny et al., 'Population Trends of the World's Monitored Seabirds, 1950–2010', *PLoS ONE* 10.6 (2015): e0129342

336 class of vertebrate: J. P. Croxall et al., 'Seabird Conservation Status, Threats and Priority Actions: A Global Assessment', *Bird Conservation International* 22.1 (March 2012): 1–34

337 smell a rat: Mailee Stanbury and James V. Briskie, 'I Smell a Rat: Can New Zealand Birds Recognize the Odor of an Invasive Mammalian Predator?', *Current Zoology* 61.1 (2015): 34–41

338 into rarity: the International Union for the Conservation of Nature collates data for threatened species in its regularly updated Red List. For fulmar, see http://www.iucnredlist.org/details/22697866/0; for the way in which seabirds are killed by trawler cables, see Benedict J. Sullivan, Timothy A. Reid and Leandro Bugoni, 'Seabird Mortality on Factory Trawlers in the Falkland Islands and Beyond', *Biological Conservation* 131 (Sept. 2006): 495–504

338 are also declining: for Atlantic puffin, see http://www.iucnredlist.org/details/22694927/0; for tufted puffin, http://www.iucnredlist.orgdetails/22694934/0; for horned puffin, http://www.iucnredlist.org/details/22694931/0

338 years after 2005: for common guillemot, see http://www.iucnredlist.org/details/22694841/0; for Brünnich's guillemot, http://www.iucnredlist.org/details/22694847/0; for razorbill, http://www.iucnredlist.org/details/22694852/0

339 begin to die: for kittiwake, see http://www.iucnredlist.org/details/
    22694497/0; Philippe M. Cury et al., 'Global Seabird Response to Forage
    Fish Depletion – One-Third for the Birds', *Science* 334.6063 (2011): 1703

339 cold northern ocean: for herring gull, see http://www.iucnredlist.org/
    details/62030608/0; for European shag, http://www.iucnredlist.org/
    details/22696894/0; for gannet, http://www.iucnredlist.org/details/
    22696657/0

340 with extinction: for Amsterdam albatross, see http://www.iucnredlist.org/
    details/22696894/0; for wandering albatross, http://www.iucnredlist.
    org/details/22698305/0; for Tristan albatross on Gough, http://www.
    iucnredlist.org/details/22728364/0; for albatrosses and petrels as a whole,
    R. A. Phillips et al., 'The Conservation Status and Priorities for Albatrosses
    and Large Petrels', *Biological Conservation* 201 (Sept. 2016): 169–83

340 seabirds towards calamity: about 400,000 seabirds die each year in gillnets.
    Ramūnas Žydelis, Cleo Small and Gemma French, 'The Incidental Catch of
    Seabirds in Gillnet Fisheries: A Global Review', *Biological Conservation* 162
    (June 2013): 76–88

343 next sixty-five years: For the future of the Atlantic puffin, see http://www.
    iucnredlist.org/details/22694927/0; for the Norwegian catastrophe, Tycho
    Anker-Nilssen, 'Status of Seabirds in Northern Norway', *British Birds* 84
    (Jan. 1991): 329–41; R. T. Barrett et al., 'The Food, Growth and Fledging
    Success of Norwegian Puffin Chicks *Fratercula arctica* in 1980–1983', *Ornis
    Scandinavica (Scandinavian Journal of Ornithology)* 18.2 (May 1987): 73–83;
    R. T. Barrett and Olav J. Runde, 'Growth and Survival of Nestling
    Kittiwakes *Rissa tridactyla* in Norway', *Ornis Scandinavica (Scandinavian
    Journal of Ornithology)* 11.3 (Dec. 1980): 228–35; Per Fauchald et al., *The
    Status and Trends of Seabirds Breeding in Norway and Svalbard*, NINA Report
    1151, Tromsø: Norwegian Institute for Nature Research, 2015

344 go as follows: for a wide-ranging summary with many references to the
    cumulative science, see Morten Frederiksen, Tycho Anker-Nilssen, Gregory
    Beaugrand and Sarah Wanless, 'Climate, Copepods and Seabirds in the
    Boreal Northeast Atlantic: Current State and Future Outlook', *Global
    Change Biology* 19.2 (2013): 364–72; Richard R. Veit and Lisa L. Manne,
    'Climate and Changing Winter Distribution of Alcids in the Northwest
    Atlantic', *Frontiers of Ecology and Evolution* 3 (April 2015): 66–74; for over-
    wintering sandeels, see Mikael van Deurs, Martin Hartvig and John Fleng
    Steffensen, 'Critical Threshold Size for Overwintering Sandeels (*Ammodytes
    marinus*)', *Marine Biology* 158.12 (2011): 2755–64

345 dead chicks: for an overview of trophic linkages, see Michael Heath,
    M. Edwards, R. Furness, J. Pinnegar and S. Wanless, (2009) 'A View from
    Above: Changing Seas, Seabirds and Food Sources', in *Marine Climate Change
    Ecosystem Linkages Report Card 2009*, online at http://strathprints.strath.
    ac.uk/29313/; H. Hátún et al., 'Large Bio-geographical Shifts in the North-

Eastern Atlantic Ocean: From the Subpolar Gyre, via Plankton, to Blue Whiting and Pilot Whales', *Progress in Oceanography* 80 (2009) 149–62

346 killed the puffins: for the oceanographic background, see Toby J. Sherwin et al., 'The Impact of Changes in North Atlantic Gyre Distribution on Water Mass Characteristics in the Rockall Trough', *ICES Journal of Marine Science* 69.5 (2012): 751–7; M. Edwards et al., 'Marine Ecosystem Response to the Atlantic Multidecadal Oscillation', *PLoS ONE* 8.2 (2013): e57212; N. Penny Holliday, 'Reversal of the 1960s to 1990s Freshening Trend in the Northeast North Atlantic and Nordic Seas', *Geophysical Research Letters* 35.3 (2008)

347 to southern Iceland: Freydís Vigfúsdóttir et al., "Large-Scale Oceanic Forces Controlling the Atlantic Puffin in S.-Iceland', *Seabird Group 10th International Conference* (2009): 64; Erpur S. Hansen, 'Sandeel Availability and Atlantic Puffin Recruitment, Mortality, and Harvest in the Westman Islands', *Seabird Group 10th International Conference* (2009): 29; Martin Edwards et al., 'Marine Ecosystem Response to the Atlantic Multidecadal Oscillation', *PLoS ONE* 8.2 (2013): e57212

348 four days earlier: Eugene J. Murphy et al., 'Climatically Driven Fluctuations in Southern Ocean Ecosystems', *Proceedings: Biological Sciences* 274 (Dec. 2007): 3057–67; Elvira S. Poloczanska et al., 'Responses of Marine Organisms to Climate Change across Oceans', *Frontiers in Marine Science*, 4 May 2016, 3: 62

348 their breeding colonies: for little auks, see David Grémillet et al., 'Arctic Warming: Non-Linear Impacts of Sea-Ice and Glacier Melt on Seabird Foraging', *Global Change Biology* 21.3 (2015): 1116–23; for seabirds creating global warming, 'Seabirds Influence Arctic Methane and Nitrous Oxide Emissions', *Eos* 94.8 (Feb. 2013): 76; for Caribbean shearwaters, Carine Precheur et al., 'Some Like It Hot: Effect of Environment on Population Dynamics of a Small Tropical Seabird in the Caribbean Region', *Ecosphere* 7.10 (2016): e01461; for sea-ice problems, Christophe Barbraud, Karine Delord and Henri Weimerskirch, 'Extreme Ecological Response of a Seabird Community to Unprecedented Sea Ice Cover', *Royal Society Open Science* 2.5 (2015): 140456

349 in those seas: Robert J. M. Crawford, 'Food, Fishing and Seabirds in the Benguela Upwelling System', *Journal of Ornithology* 148.2 (2007): 253–60

349 ages of loss: for the general catastrophe and the current wave of extinctions, see James A. Estes et al., 'Trophic Downgrading of Planet Earth', *Science* 333.6040 (2011): 301–6; for a variety of outcomes, see Keith C. Hamer, 'The Search for Winners and Losers in a Sea of Climate Change', *Ibis* 152.1 (2010): 3–5; for a more optimistic (but possibly dated) picture, see Marine Grandgeorge et al., 'Resilience of the British and Irish Seabird Community in the Twentieth Century', *Aquatic Biology* 4.2 (2008): 187–99

350 gates of extinction: for seabird plasticity, see David Grémillet and Anne Charmantier, 'Shifts in Phenotypic Plasticity Constrain the Value of Seabirds

as Ecological Indicators of Marine Ecosystems', *Ecological Applications* 20.6 (2010): 1498–1503; Christophe Barbraud et al., 'Effects of Climate Change and Fisheries Bycatch on Southern Ocean Seabirds: A Review', *Marine Ecology Progress Series* 454 (2012): 285–307; C. Barbraud, G. N. Tuck, R. Thomson, K. Delord and H. Weimerskirch, 'Fisheries Bycatch as an Inadvertent Human-Induced Evolutionary Mechanism', *PLoS ONE* 8.4 (2013): e60353; Geoffrey N. Tuck et al., 'An Integrated Assessment Model of Seabird Population Dynamics: Can Individual Heterogeneity in Susceptibility to Fishing Explain Abundance Trends in Crozet Wandering Albatross?', *Journal of Applied Ecology* 52.4 (2015): 950–9; Jean Rouziès, 'Le Chercheur de Chizé et le paradoxe des albatros', *La Nouvelle République*, 28 April 2013

353 before rescue came: for the Shiant Isles Recovery Project see http://www.rspb.org.uk/our-work/conservation/shiantisles/work/index.aspx and its associated blogs and newsletters; the initial survey of British islands selecting the Shiants is in Norman Ratcliffe et al., 'How to Prioritize Rat Management for the Benefit of Petrels: A Case Study of the UK, Channel Islands and Isle of Man', *Ibis* 151.4 (2009): 699–708, online at http://www.pacificinvasivesinitiative.org

Acknowledgements

This book owes debts everywhere it touches.

First, on the Shiant Isles for education and illumination to all those involved in the Shiant Seabird Recovery Project, especially George Campbell, Phil Taylor, Thomas Churchyard, Biz Bell, Robin Reid, Davide Scridel, Emily Scragg, Laura Robertson, Niall Currie and Charlie Main. Ellie Owen was instantly helpful in giving me access to map data from the RSPB's tracking projects.

At Scottish Natural Heritage, thanks, as ever, to David MacLennan, Roddy MacMinn and Andy Douse for being consistently helpful and broadminded in looking after the islands and their birds.

No year on the Shiants would be complete without the revelatory and high-intensity fortnight in which Jim Lennon and the Shiants Auk Ringing Group manage, year after year, to ring their 2,000 birds.

On Scalpay a thousand thanks to Roddy MacSween, Rachel and Donald MacSween, Effie Park and Morag Macleod for all their help in untold ways; on Lewis, to Margaret, Charles and Johannes Engebretsen, to Donald 'Dolishan' Macleod, to Donald Murray and to Ian Stephen for an unforgettable day on his beloved *Broad Bay*, me talking and helming, him muttering and lying in the bilges.

Several of the journeys in this book, to Iceland, the Skelligs, the Faeroes, Bass Rock, Orkney, the Falklands and South Georgia, were made when filming for BBC4. For wonderful and eye-opening times, all thanks to the executive producers at Keofilms, Andrew Palmer and Will Anderson, the directors Tom Beard and James Nutt and the indefatigably resourceful and life-enhancing assistant producers Adam Scott and Scott Ward. None of it would have been possible without the work of the great team at Keo North: Maddy Allen, Justine Leahy, Ciara Spankie and Ryan Small.

Every seabird scientist I have asked for help and guidance has given freely and generously of their time and advice. This book would have been unthinkable without the thousands of years of work which the seabird and ocean science community has dedicated to understanding the birds, often in the most demanding and hostile of circumstances. Among those who have freely shared their research and discoveries are: Douglas Russell, Curator of Eggs at the Natural History Museum in Tring, Ian Mitchell, Heather McLannahan, Robert Furness, Erpur Hansen, Aevar Petersen, Ewan W. J. Edwards, Paul Thompson, John Croxall, Tycho Anker-Nilssen, Tim Guilford, Michael Brooke, Martin Heubeck, Samantha Patrick, Penny Holliday, Mike Harris and Kenny Taylor.

Above all, I would like to thank Sarah Wanless and Ewan Wakefield for their generosity, time, expertise, vital spirit and encouragement while I was writing this book, for whose contents of course they bear no responsibility. Sarah's work with Mike Harris over many years on the Isle of May has moulded the modern understanding of the seabirds she loves. Ewan is part of the new generation of seabird scientists on much of whose work this book has relied.

As fellow socio-bio-culturo-ornitho-oceanic historians, I would also like to thank Melanie Challenger, Graham Barwell,

John Baldwin, Claire Preston, Claire Williams, David Gange and Patricia Lysaght for all their help and insights.

In the Faeroes Bjørn Patursson; in Reykjavik Jóhann Sigurjónsson; on Grímsey, Siggi and Harpa Henningson; and on Alsey, Grímur Gíslason, Pálli Guðmundsson, Huginn Sær Grímsson, Sigurgeir Jónasson, Guðlaugur Sigurgeirsson and Diddi Sigursteinn all welcomed me into their seabird world.

I would like to thank my wonderful family and my friends Katrina Porteous, Tim Dee, Claire Spottiswoode, Michael Longley, Matt Ridley, Matthew Rice, Jim Richardson, Alexandra Chaldecott, Tom Hammick and Belinda Giles, for the most precious thing they have all given me: solid, true encouragement.

To Lisbet Rausing a particular thank-you for her campaign to broaden open electronic access to the scientific research which would otherwise remain hidden behind pay-walls. This book could not have been written without the changes she has argued for, worked for and engineered.

To Mike Buckland, many thanks for describing so cheerfully and generously his terrifying encounter with a gannet.

As always, I would like to thank Georgina Capel and Zoe Pagnamenta for their sustenance and friendship over the years; to Peter James, Lottie Fyfe, Arabella Pike, Susan Watt and Jack Macrae all thanks for their editorial brilliance and firmness, without which I would still be wandering more lost than a shearwater somewhere off the Azores with corks in its nose.

Kate Boxer's wonderful pictures of seabirds adorn and shape this book. More than anyone I know, she can paint the inner sensibility of creatures, not merely the body-form but its motivating spirit. In 1909 Jakob von Uexküll was the first to identify what he called the *Innenwelt* of animals, that particular interiority derived from the animal's grasp of the world that surrounds it. All animals meet the world in their own way and that unique understanding of their surrounding world, their *Umwelt*, shapes the beings they are. *Umwelt* makes *Innenwelt*. Uexküll's powerful idea has rippled down through animal science in the century since he wrote: to Konrad Lorenz, to Niko Tinbergen, to the Oxford school, of which Bryan Nelson, the gannet scientist, was a member, and on from him to his pupil Sarah Wanless and from her to Francis Daunt and many others. It is the central, glowing thread of modern seabird science, the principle Harper Lee put in the mouth of Atticus Finch in *To Kill A Mockingbird*: understand the other 'from his point of view … climb into his skin and walk around in it'. The future of the seabirds – and perhaps of all living things – depends on that becoming the universal understanding. 'Cow parsley has a language of its own,' Ali Smith wrote in *Autumn*, her novel published in 2016. Yes: cow parsley, puffins, sandeels, gannets, plankton, the great wandering albatrosses, every living thing in a multiple Babel of radiant interiorities. Miraculously, Kate Boxer has depicted it in these birds and it is impossible to thank her enough for it.

Illustration Credits

Index